U0897136

Shopping Experience

Shopping Experience

体验空间

于　萍　编
常文心　译

辽宁科学技术出版社

Contents
目录

Electronic Products
电子类

Toys
玩具类

Index
索引

Aurelio Vázquez Durán
DIN Interiorismo

1. How do you understand the concept of the Experience Store?

A: The idea was to change the traditional "furniture store" and transform it into a space where the visitor gets the feel of a real ambiance through all the areas and sets in the store. Visiting the store is not just a mere shopping trip; the customers will enjoy just walking and picturing the different scenes displayed -such as: dinning rooms, living rooms, bedrooms, etc.– as part of their own homes and eventually buy all the set or just select an accessory or two.

There are no corridors, so the customers may browse through all the areas just noticing the change of section because of the different products in display. Indirect lighting to emphasise the areas in combination with natural light adds the special atmosphere to every area. The ceilings have a metallic structure so different elements may be hung to enhance the displays and change according to the seasons or special promotions.

2. What do you think is the most important for an experience store?

A: The most important thing when designing an experience store is to forget about selling products and to centre in the client's experience, to determine how is he going to enjoy and live it.

3. How to design the atmosphere of an experience store?

A: The atmosphere of an experience store has to emerge from livable, mouth watering and desirable spaces that give solutions and surprises with permanent changes. An atmosphere like this will make the customer want to come back frequently because he feels comfortable and most of all he knows it is the right place to satisfy his interior design needs.

4. How to display the cultural element of the products?

A: When talking about cultural elements we have to understand first that globalisation is a fact, so we have to be very careful with the display in order to create a tasteful display. The main keys to achieve this are to focus on truth and credibility, the rest is left to creativity.

5. How to divide the space in the store to reach the best effect?

A: To divide the space the best thing is to have the fewer hard walls possible. Open spaces will generate a more flexible and dynamic space, that will allow all the elements to interact and disappear as needed.

6. What kind of added value does the design of the experience store bring for the products?

A: The design of the experience store is definitely the wrapper of the product. The correct atmosphere will generate the perfect ambiance, hence the product will catch the eye better. Sometimes just a spot of light on top of the product will make it stand out from the rest when the interior design is based on the experience.

7. What's your design technique (like style, material, proportion and so on) to make the display part stand out?

A: My design technique is to balance all the elements in order to achieve the results that the store needs. Even though the main target of the store is to sell products, it is very important that the displays do not take over its personality. In my experience you have three main levels: the client, the product and the store, and the three have to be taken into consideration to obtain a correct result.

奥雷里奥·瓦斯克·杜兰
DIN室内设计公司

1.您怎样理解体验店的概念?

这个概念的主旨是要改变传统的家具店，使其成为一个能让顾客通过店内布局和设置感受到真正的体验氛围的空间。逛这种商店不仅仅是单纯的购物之旅，客人们还将会享受这种在店内自在随意的游走，观赏到许多不同的精心设计的、就仿佛他们家中的某部分的场景，例如：餐厅、起居室、卧室等等，最后在舒适的体验中挑选自己喜欢的一套或者一两件商品。

店内空间里面没有固定走廊，这样顾客们可以在整个室内随意的闲逛，通过展示出来的不同产品来注意到格局的变换。人工光源和自然光线的融合会为各个区域增添独特的氛围。天花板采用金属结构，这样悬挂在上方的装饰元素可以更好的突出展品，并且可以很方便的根据不同的季节或者特殊的目的进行变换。

2.您认为体验店设计的最重要的元素是什么?

设计一家体验店最重要的一点是要忘记你是在销售商品，而是要更关注客人们的体验感受，明确客人们会怎样享受和体验它。

3.怎样营造体验店内的空间氛围?

体验店的氛围必须是温馨居家、秀色可餐并且让人充满想象与渴望的，这个空间会不断变化，给客人以引导和惊喜。这样的一个空间会让客人常常想要光顾，因为他觉得在店里很舒适很放松，而且最重要的是他知道他对室内设计的需求会在这里得到很好的满足。

4.怎样展示商品的文化元素?

当我们谈到文化元素这个话题时我们必须了解全球化现象这一事实。所以我们必须小心确保文化元素展示的目的是营造一个有品位的空间。要达到这一点的关键是着眼于事实和现状，其余的就留给创造力的发挥。

5.怎样划分空间来达到最好的效果?

划分空间最好的就是使用尽可能少的墙。一个开放的空间会创造出更灵活更有活力的氛围，有利于人们根据需求对各个元素进行融合和分离。

6.体验店的设计会给商品带来什么附加价值?

体验店的设计绝对是商品最好的宣传。正确的体验店设计会创造出适合商品的完美的氛围，因此商品也会更加吸引顾客的眼球。如果室内设计是着眼于体验的话，那么有时候商品上方的一束灯光就会使商品凸显于其他的周围环境。

7.您突出展示部分的设计手法是什么（例如设计风格，使用材料，空间比例等等）?

我的设计手法是平衡所有的设计元素以达到商店所需的最佳效果。即使出售店内的商品是商店经营的主要目的，但是不要让展示的商品夺走了商店的个性和特点也是非常重要的。在我看来设计目标有三个层次：第一层是顾客，第二层是商品和店面，第三层最重要，就是要考虑达到正确并完美的效果。

Ilaria Marelli
Ilaria marelli studio

Each of my project, in interior, and in exhibit design, starts from a strong and clear idea, that unify the space and make it immersive for customers, a space where to experience emotions and beauty.

I feel I am a sort of "storyteller" through my design, and I see that people always understand a project if they perceive that there is a story behind that.

The experiencing centre in my opinion, is always a story displayed in a space that mixes elements and reminds previous experiences with something surprising, poetic and beautiful which makes people stay more, and usually share it with friends.

The starting point is always a mix of cultural influences, intuition, projection into the future, and a careful listening to people's desires.

This is an important point in my design...at the basic level I work to obtain a friendly experience for users: they must be guided to understand the path – the display of product, etc, but more than that I take care that users get fascinated by the general mood of the space.

I always add something "magic" in my interior design. It could be an unusual element of display (such as real trees, out of scale furniture, boat shells...). Plus I always give a special attention to light (very bright – or very dark – to change the mood of people coming from outside), and to a general welcoming atmosphere.

Usually I try to surprise the customer directly at the entrance, so that he feels entering in something different. The door is the gateway, and then I add some touch of "magic" as heart to discover in every centre of different displays...If there are more brands, each one will have its own peculiar "emotional display"; the same if the space is divided into different categories or themes, each one must have its own personality.

Then the specific details are to be defined according to the identity of the space and its purpose (selling, educational...), and also with the director of the space, since I have a lot of respect for their knowledge in everyday use of my design and their strategy to display products in a better way.

伊拉莉亚·马瑞莉
伊拉莉亚·马瑞莉设计事务所

我设计的每一个项目，无论是室内项目还是展览类项目，都是从一个强烈而又清晰的思路开始的，即融合并统一空间中的元素，使顾客沉醉其中，让其成为一个能感受到情感与美感的空间。

在设计的过程中，相对于设计师的身份来说，我感觉我更像是一个故事讲述者，而且我发现人们在感受到设计的背后存在着一个故事的时候，总是能很好的理解这个设计项目。

在我看来，体验店就像是在空间里展示着的一个美丽的故事，它融合了许多元素，用令人惊喜的、诗情画意的和美丽优雅的东西唤起人们珍贵的回忆，让人们在店内流连忘返，并且愿意与朋友分享这一切。

设计的起点是把文化影响、感官直觉和未来构想糅杂在一起，并且要用心倾听人们的需求与渴望。

这些都是我在设计中注重的观点。在基本的层面上我是要为顾客营造一个温馨美好的体验感受：他们会被引领着找到路线，领会商品展示的内涵等等，但我更注重的是顾客要被空间中散发出来的整体气氛所吸引、所倾倒。

我在室内设计中总会添加一些“魔法”在里面，它可以是在展示中不同寻常的元素（例如真实的树木，不合比例的家具，船的外壳等等）。此外，我还特别注重灯光的使用（明亮一黑暗，可以改变进入店铺中的客人的心情），并且灯光还可以营造很好的欢迎客人的氛围。

通常，我会在店铺的入口处就给顾客设置一些惊喜，这样他们会觉得自己来到了一个不同寻常的地方， 通道就是大门，我在每个不同展示区的中心都添加了一些小“魔法”让顾客自己去发现……如果店内有许多品牌的商品，那么每个品牌都会有自己独特的“情感展示”，同一空间被分隔成不同的门类和主题的时候也是同样如此，每一个部分都要有自己特有的个性。

除此之外，店内特殊细节的设计要根据空间的特点和它的目的性来界定（如销售、教育等等），而且一定要和店铺的主人进行良好的沟通，因为我对店主在怎样利用我的设计和他们怎样更好的来展示自己的商品的经验和技巧一直怀有深深的尊敬。

LABscape
Tecla Tangorra & Robert Ivanov + Cecilia Bernasconi de Luca

1. How do you understand the concept of the Experience Store?

It is to put the customer in an unusual ambiance, to create experience in an unusual place, to find the right product, to create an experience in the act of purchase, not only an aesthetical space, or innovative, but also create a range of services that complete the promenade.

2. What do you think is the most important for an experience store?

Curiosity - Create a place where the product and the space are merging together, adding technology and interaction with the customer. The experience is an act that calls for the five senses and memory. If in a shop you have at least two senses stimulated, then it has partly succeeded in creating experience.

3. How to design the atmosphere of an experience store?

There isn't a universal receipt. But we very often see how organised system in nature works then we try to recreate it for the project incorporating the programme and all the constrains. In the first process of the design it will be the concept to find in correlation with the services. The second process is to adapt the concept to the functions and purposes.

4. What kind of added value does the design of the experience store bring for the products?

The product is no longer shown as a stand-alone object but is taking part of the overall scenography, which creates the complete experience.

When the product and the space are completely working together it's making a new way to see the product, because the product is taking all of his meaning inside the space. The values of the product are completely

linked with the experience that the customer has while is in the store. It makes the client closer to the situation of choosing the right product for him.

LAB scape 建筑事务所
特卡拉·唐戈拉 & 罗伯特·伊万诺夫 + 塞西莉亚·贝纳斯科尼·卢卡

1.您怎样理解体验店这一概念?

我认为，体验店能让顾客置身于一个与众不同的环境，在这个与众不同的地方形成一种经历，它会引导顾客寻找到正确的合适的产品，并使他们在商品交易中得到一种愉悦的体验。体验店设计不仅仅是注重空间的美感和创新，还要设计出一系列的细致服务来完善这个自由的空间。

2.您认为体验店设计的最重要的元素是什么?

新奇——创造一个商品与空间能够很好融合在一起的地点，添加科技元素，并与顾客形成互动。体验是一种能够唤起人们五种感官和形成回忆的一种活动，如果你在一家商店里被唤起了至少两种感官的共鸣并且创造了一种印象，那么可以说这家店已经在创造体验性空间中获得了成功。

3.怎样营造体验店的氛围?

人与人的品位、喜好各不相同。我们经常看到有机系统在大自然中是怎样运作的，那么，我们也可以试着融合计划和限制为项目重现一个有机体系。设计的第一步是要考虑到与服务相关的方方面面。第二步是注意着眼店内功能性与目的性的设计构想。

4. 体验店的设计会给商品带来什么附加价值?

体验店内的商品将不再只是单独展出的一个个体，而是已经与周围环境融为一体，共同创造出一个给顾客舒适体验的空间。
当商品与店内空间完美地融合在一起时，人们会用全新的方式去审视这些商品，因为商品本身就在诉说着空间的内在含义。商品的价值完全与顾客在店内停留时感受到的美好体验联系在一起。这样更有利于顾客切身体会并挑选到真正适合自己的产品。

1

Calle Veinte

二十街装饰店

Location:
Mexico City, Mexico

Designer:
DCPP Arquitectos

Photographer:
Onnis Luque

Completion date:
2009

项目地点：
墨西哥 墨西哥城

设计师：
DCPP建筑事务所

摄影师：
奥尼斯·卢克

完成时间：
2009

Calle Veinte is a decoration store situated in a shopping centre on the first floor facing the street. The project is located in high vehicular traffic street and responds to it as an urban sideboard. The ground floor works as this great sideboard to the exterior, at eye's elevation and at the vehicle's speed you can appreciate the ceiling and that is why it's one of the subjects that the project pretends to emphasise. The ceiling is made of wooden beams in repetition which give the space a sense of depth and height as well as break its horizontality and provide a certain rhythm.

The sense of cleanliness in the inside, allows the furniture and fabrics to give the space colour and life, making evident only the structure, which contrasts with the white colour of the walls. The furniture and fabrics can naturally be the decoration; thus no extra complicated decorations are needed. The goods can be highlighted.

The materials were used in their natural condition, the marble on the floor, the natural wood on the ceiling and the transparent glass on the walls, leaving all the prominence to the furniture. In this sense, nature and human are combined together perfectly.

二十街装饰店位于一个购物中心二楼临街的一侧，楼下车水马龙，十分繁华。一楼宛如一个都市看板，可以看到室外的风景，而天花板也是项目的设计重点之一。天花板由一系列的木条组成，增添了空间的深度和高度，打破了水平设计，也形成了一种韵律感。

室内空间的简洁感让家具和织物焕发出色彩和活力，与白色的墙壁形成了鲜明的对比。家具和织物本身就是一种装饰，因此，空间里并不需要更多复杂的装饰，商品被很好地突出了。

设计的材料全部采用自然形态，如地面的大理石、天花板的实木、墙壁上的透明玻璃。这些材料都为家具让步，凸显家具的品质和特点。这样一来，自然和人性完美地结合在一起。

2

1. A concise and clean interior plan
2. The busy traffic outside seen through the windows makes a contrast with the quiet interior atmosphere
3. Decorations of retro style
4. Taking the white wall as a backdrop, the fabrics look more colourful
5. The open space provides customers with more comfortable shopping experience
6. The ceiling and floor are made of natural wood and stone

1.简洁干练的室内布局
2.透过玻璃窗可以看到外面的车水马龙，与室内的静谧形成对比
3.复古风格的家饰
4.白墙映衬下的织物显得更加五彩缤纷
5.宽敞的空间让顾客有更舒适的购物体验
6.天花板和地板都采用天然木材和石材

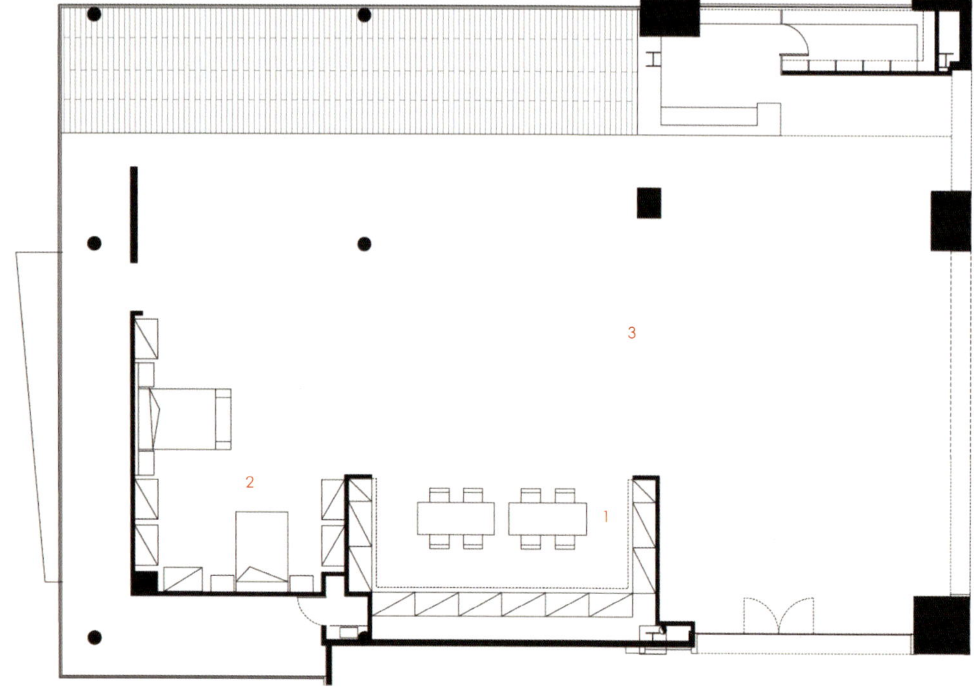

1. Salon 1.沙龙
2. Cashier desk 2.收银台
3. Exhibition area 3.展示区

Estilo Modernidad Seren

ALHAMBRA
COJINES • BLANCOS • TOALLAS
CORTINAS • TAPICERIA • SOFAS

4

Comodidad Acogedor Carácter Calidez Equilibrio
5
6

Casa Palacio

宫殿之家

Location:
Mexico City, Mexico

Designer:
DIN Interiorismo, Aurelio Vázquez Durán

Photographer:
Casa Palacio

Completion date:
2007

项目地点：
墨西哥 墨西哥城

设计师：
DIN室内设计，奥雷里奥·瓦斯克·杜兰

摄影师：
宫殿之家

完成时间：
2007

This project was a challenge of coordination, consultancy and stage design for the first store specialised in home furniture and accessories of the Grupo El Palacio de Hierro, one of the most important retail stores in Mexico. The main target was to develop an interior design with the commitment of offering a space where the most important aspect is to feel the atmosphere where lifestyle is the essence.

While browsing through Casa Palacio the customers discover that each space has a particular lighting, aroma and music, all selected specifically for each area. The transition from one to the other is very soft, as all the areas may be reached from several directions, opening a wide variety of views and different routes to discover the whole project.

The store was designed with experiences more than products in mind, in the interior of a round shape building of 6,000 square metres. The two levels are connected with electric staircases under the natural light of a magnificent dome.

项目是墨西哥零售大亨耶罗宫殿集团的首家家居用品店，设计师承担了合作、咨询和展示设计工作。项目的主要目标是打造一个生活风格十足的空间。

在宫殿之家里，顾客会发现每个区域都有自己独特的灯光、芳香和音乐。区域之间的过渡十分柔和自然，每个区域都能从好几个不同的方向进入，拥有多方位的风景，为整个项目提供了不同路径。

商店位于一个圆形建筑中，总面积为6,000平方米，力求为顾客提供完美的购物体验。两层楼之间由电梯相连，巨大的华盖能够提供自然采光。

1. Furniture is arranged into different areas in this round building
2. The colourful towels add lively elements to the interior
3. A showroom of living room
4. A warm and cosy European-style dining area
5. The hung lamp has a distinguishing feature
6. A lovely baby room
7. Different areas transit naturally and softly
8. A classic and elegant bedroom

1.圆形建筑内家具分区布局
2.彩色的毛巾增添了室内的活力
3.客厅摆设样例
4.温馨的欧式餐区布局
5.悬垂的吊灯充满特色
6.婴儿房整洁可爱
7.各个区域过渡柔和自然
8.卧房布局典雅沉稳

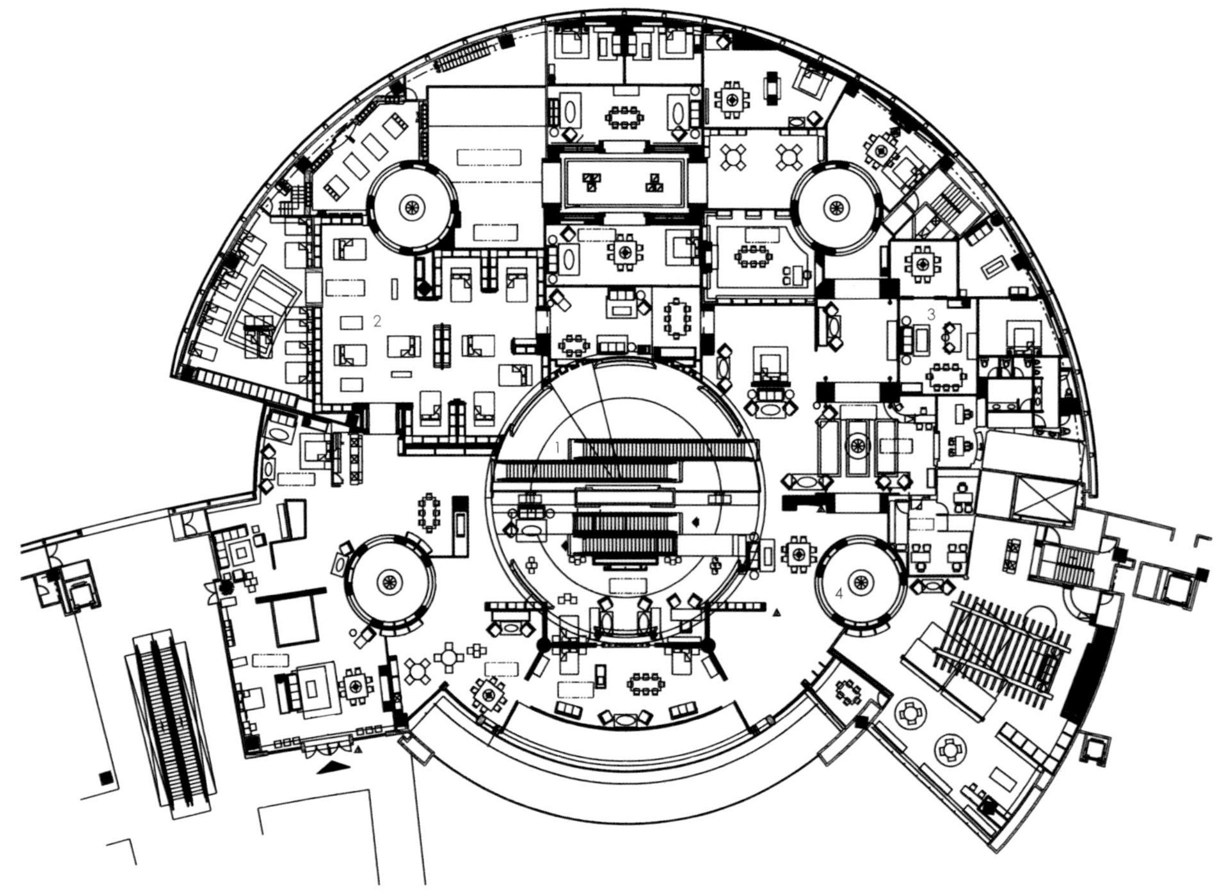

1. Electric staircases
2. Bedroom display area
3. Living room display area
4. Pillar

1.电梯
2.卧室展示区
3.起居室展示区
4.柱子

3

4

5

6

7

8

1

Davis + Warshow Showroom

戴维斯+华修展示厅

Location:
New York, USA

Designer:
Ronnette Riley Architect

Photographer:
Esto Peter Aaron

Completion date:
2007

项目地点：
美国 纽约

设计师：
罗尼特·瑞利建筑事务所

摄影师：
艾斯托·皮特·阿伦

完成时间：
2007

Davis + Warshow, New York's oldest and largest plumbing fixture distributor, hired Ronnette Riley Architect (RRA) to design a 670-square-metre showroom and event space to became their flagship store. They were looking for a retail feel, focusing not on brand recognition, as is the industry standard, but on the fixtures themselves. RRA strived to make every aspect of the showroom adjustable and interchangeable in order to meet the needs of the ever changing fashion of fixtures and to allow for the flexibility needed to accommodate various size groups for events.

Davis + Warshow took a stand on presenting their fixtures without the use of running water but its inevitable presence in the showroom was essential. RRA enlisted subtle hints of movement and flowing materials to create the illusion of water. The first step into the showroom is onto a brilliant, sienna orange, rubber mat. This squishy flooring, which was originally created as a writing surface, slides up and over the reception desk. As the guest approaches the receptionist, they will pass the rubber as it peels off the main walk into a curving bench and magazine shelf and a waterfalling wall of rubber dotted with honey-coloured Chroma Resin blocks. The colours and materials were carefully selected to capture the excitement and vibrancy of the SOHO neighbourhood.

戴维斯+华修是纽约最大、历史最悠久的卫浴设备经销商，罗尼特·瑞利建筑事务所为他们所设计的旗舰店总面积为670平方米，包括展示厅和举办活动的空间。戴维斯+华修公司追求一种零售氛围，这种氛围不是以品牌辨识度和产业标准为中心，而是更注重卫浴设备本身。罗尼特·瑞利建筑事务所力求让展示厅的每部分都能灵活调整，不仅要满足不断变化的卫浴时尚，还要能灵活调整，举办大小活动。

戴维斯+华修坚持不在展示厅内采用流水，但是水又是卫浴设备必不可少的元素，于是设计师运用流动的材料打造了水的错觉。进入展示厅，便是土黄色的发光橡胶垫。这块粘软的地面一直向上滑到前台的桌面上。当顾客走向前台时，他们会经过这块橡胶垫，而橡胶垫又卷曲成为长椅、杂志架、橡胶墙等。橡胶墙面上点缀着蜂蜜色的显像管方块。色彩和材料的选择充分反映了周边SOHO区的动感与活力。

展品被放置在36个可移动的展位里，展位设计采用1:2的比例，便于移走以举办活动或是重新设计展示厅。随着展示产品的改变，展位上的石板可以被移开或重置。为了适应不断变换的展示内容，设计师设计了一个可由前台控制的灯光系统，可以瞬间将零售展厅转变为夜间展览会，甚至是夸张的舞会。在举行重大活动时，展位可以被移到销售区，藏在一面巨大的滑动面板后面，灯光将只打在被保留下来的展位上。

销售区与展示厅之间由一扇低矮的半透明树脂隔断隔开，便于销售助理监控展示厅里顾客的情况，而他们自己则可以隐藏在隔断后面。为了适应销售助理人数的增减，水平工作区也采取了滑动隔断和档案柜。

2

1. A general view of the showroom
2. The wall of rubber is dotted with honey-coloured Chroma Resin blocks, providing the space with energy and dynamic
3. The sales area is separated from the showroom by low semi-opaque resin dividers
4. The display of a bathroom

1.展示厅纵览
2.橡胶墙面上点缀着蜂蜜色的显像管方块，使空间充满动感与活力
3.销售区与展示厅之间由半透明树脂隔断隔开
4.浴室展示与体验

Displays are set in thirty-six moveable "Pods", designed on a 1:2 ratio, which allows them the utmost flexibility when rearranging for events or to redesign the showroom. All displays are designed with drop in panels of stone which can be lifted out, and replaced, as the products change. To accommodate the ever changing and endless display options RRA designed a lighting system which allows the receptionist to choose between several pre-set lighting schemes, instantly bringing the showroom from a retail focus to an evening presentation space, or a dramatic cocktail party. For a large event, the moveable Pods are relocated into the sales area, hidden by a large sliding panel, and the lighting scheme will be switched to place lighting emphasis only on the remaining pods.

The sales area is separated from the showroom by low semi-opaque resin dividers that enable sales associates to monitor customers in the showroom, while they themselves are obscured by the reeded panels. As the need for sales associates expands and contracts so too does the horizontal work space with a continuously mounted rail that provides an axis for sliding dividers and filing cabinets.

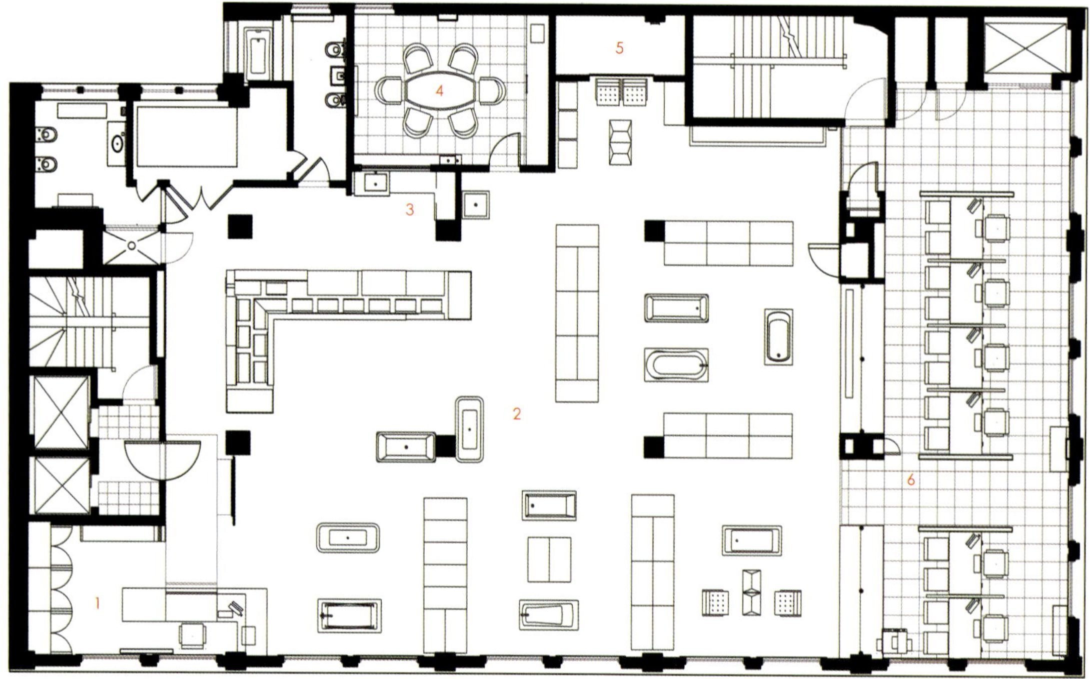

1. Reception 1.前台
2. Showroom 2.展览室
3. Pantry 3.备餐室
4. Conference 4.会议室
5. Storage 5.储藏室
6. Sales 6.销售区

3
4

1

Italia Shop-in-Shop

Italia店中店

Location:
Boston, USA

Designer:
B&B Italia SpA

Completion date:
2007

项目地点：
美国 波士顿

设计师：
B&B意大利SpA设计公司

完成时间：
2007

Over the past few years, the activity developed by B&B Italia has been distinguished by its dynamism and creativity. Thanks to the company success in identifying previously unexplored market segments, B&B Italia is now expanding into a new sector and creating an outdoor collection. With the outdoor collection, B&B Italia's innovation and design goes outdoors, providing leisurely moments in a comfortable, aesthetically pleasing setting.

Brand-new is the collection Canasta, designed by Patricia Urquiola, born thanks to the spirit of experimentation and exploration that is the hallmark of the brand. The inspiration comes from interlacing Vienna straw, which is totally reinterpreted in a contemporary way and treated with polyethylene for enhanced sturdiness, lightness and durability over time. The perfect position for these beautiful, large designs is at the pool or amidst the greenery in the garden.

With their inviting, almost anatomic shapes available in different sizes and heights, they stand out for their originality and superior quality. The collection presented here is the first steps of a project that will continue to be developed.

近几年来，B&B Italia的独特活动尽显其品牌的活力与创意。由于公司对未开发的市场进行了细分，B&B Italia现在新开发了一个户外系列。在户外系列中，B&B Italia创意和设计走到了户外，提供了闲适、美观、愉悦的家具作品。

帕特里夏・乌其欧拉的凯纳斯特户外系列诞生于该品牌的试验和探索精神，其设计灵感来自于交织的维也纳稻草。经过现代诠释，稻草由聚乙烯代替，增强了韧性、轻之感和耐久度。这些美观的巨型设计最适合摆放在游泳池旁或是花园之中。

这些讨人喜欢的家具尺寸、大小不一，具有原创性和高品质。这一系列的用品将会进行进一步的开发。

1. It is a clean and elegant space
2. The warm tone of the lighting creates a comfortable experience
3. The white backdrop is bright and clear
4. The carpet and sofa match well and have a rich texture
5. Customers can enjoy the cosy furniture in display
6. The lamp and background wall level up the modern sense

1.空间摆放整洁高雅
2.灯光色调温暖，营造了舒适的体验氛围
3.整体的白色背景洁净明亮
4.地毯和沙发搭配和谐，很有质感
5.顾客可以舒适的体验室内陈列的家具摆设
6.吊灯和背景墙提升了室内的现代感

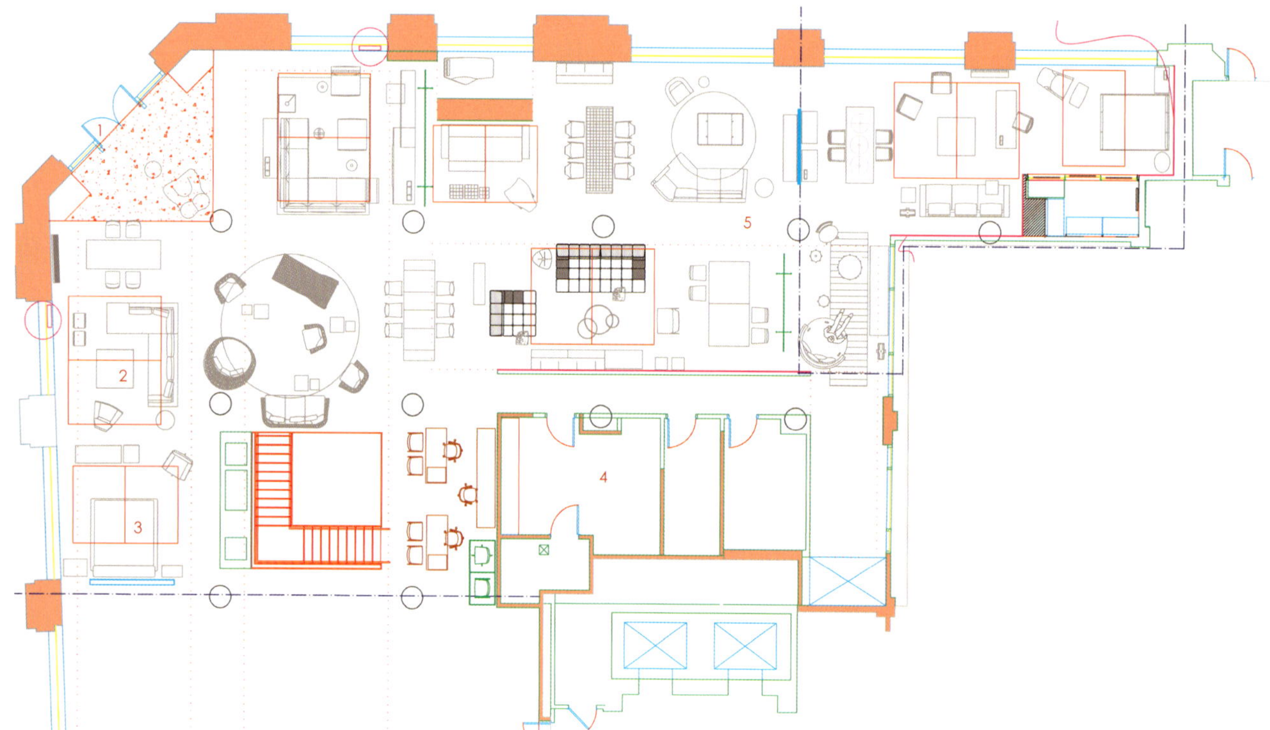

1. Entrance 1.入口
2. Living room area 2.起居室区
3. Bedroom area 3.卧室区
4. Storage 4.仓库
5. Main display area 5.主展示区

4

5

22

1

Maxalto Store
麦克萨多家居用品店

Location:
Paris, France

Designer:
Antonio Virga

Completion date:
2008

项目地点：
法国 巴黎

设计师：
安东尼奥·维嘉

完成时间：
2008

Since 2001, Maxalto, a symbol of "Made in Italy" excellence, has been located on 43 Rue du Bac, in the heart of the Saint-Germain quarter. A leading name in the world of high-quality furniture, Maxalto uses traditional furniture making techniques, modern technologies and prestigious materials, predominantly wood, to create furnishing collections with timeless and precious lines.

The main decorative tendencies of the 20th century, enriched with ideas from the best contemporary designers, are the driving force behind their creations. Since 1975 Maxalto has achieved the highest degree of excellence, as implied by the name itself.

In September 2008, Maxalto Paris inaugurated an addition that expands on Antonio Citterio's concept of a refined apartment, the underlying design concept of the original store. This architect and designer, in collaboration with the architect Antonio Virga, were the authors of this expansion built below the current showroom.

The dark wood floors and the soft atmosphere are in perfect harmony with the 280-square-metre pre-existing space. The latest Maxalto collections are displaying in the boudoir atmosphere of this extraordinary space.

麦克萨多是意大利知名品牌，2001年起，麦克萨多家居用品店登陆巴黎圣日耳曼区的鲁德贝克街。作为享誉全球的高品质家居品牌，麦克萨多采用传统家具制作工艺，运用现代科技和珍贵材料（主要为木材），打造了线条精炼的经典家具。

充满了优秀现代设计师理念的20世纪主流装饰理念是他们创造的动力。自1975年以来，麦克萨多品牌作品的品质达到了极致，与该品牌的名字（意为“最高”）十分相配。

2008年9月，麦克萨多巴黎店进行了扩张，新增了一个融入安东尼奥·奇特里奥设计理念的精品公寓式区域。安东尼奥·奇特里奥与建筑师安东尼奥·维嘉共同合作，在原有展示厅下面新设计了一个展示厅。

深色木地板和柔和的氛围在这个280平方米的空间呈现出完美的和谐状态。麦克萨多的新品家具将被展示在这个类似会客室的非凡空间。

展示厅经理布里吉特·希尔维拉将运用425平方米（包括原有展厅面积）的销售空间与顾客们共同分享麦克萨多家居体验。这个更大的销售空间将提供该品牌更多的咨询，并且会一直保留该店一贯的私密、精致的氛围。

EXIT

1. The yellow lighting adds warmth to the interior space
2. Different areas transit naturally and softly
3. Walls of red bricks are simple and unique
4. The dark wood floor combines with the interior style harmoniously
5. The cosy sofa and dignified carpet match perfectly
6. The droplights and dinning tables look decent and sophisticated

1.黄色的灯光增添了室内温暖的氛围
2.各个空间过渡柔和自然
3.红砖墙面古朴另类
4.深色的木质地板与室内风格和谐融为一体
5.舒适的沙发与厚重的地毯搭配完美
6.吊灯与餐桌风格深沉内敛

The showroom manager, Brigitte Silvera, uses the 425-square-metre selling space to share the prestigious Maxalto experience with clients. This larger selling space offers greater visibility to the creative world of the brand, without restoring to excess and without altering the intimate and refined atmosphere that Brigitte Silvera has maintained since the store first opened.

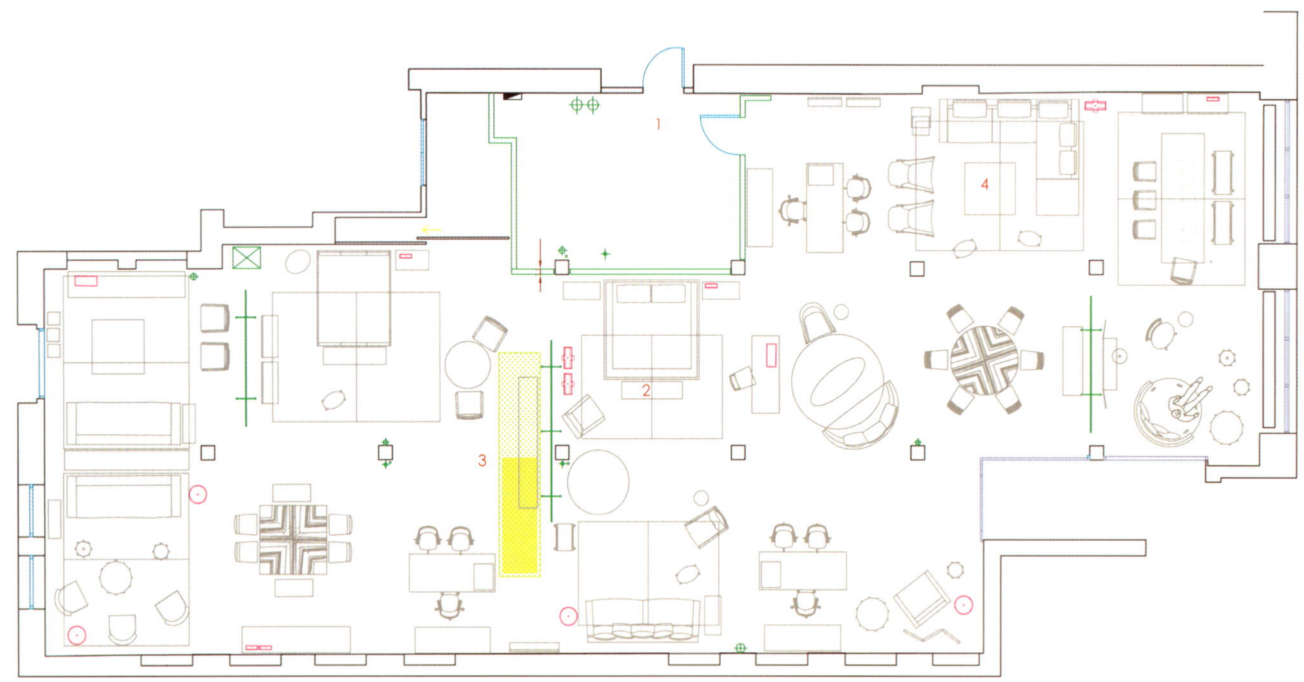

1. Entrance 1.入口
2. Bedroom area 2.卧室区
3. Shelf 3.货架
4. Living room area 4.起居室区

4

5

1

Novo Ambiente Store

诺瓦环境商店

Location:
Rio de Janeiro, Brazil

Designer:
Ivan Rezende Architecture

Photographer:
MCA Estudio

Completion date:
2008

项目地点：
巴西 里约热内卢

设计师：
伊万·理赞德建筑事务所

摄影师：
MCA工作室

完成时间：
2008

Located in Rio de Janeiro, inside a shopping centre specialising in the home and home decoration market, the shop occupies a total area of 970-square-metre covering showrooms and other facilities such as meeting room, store, toilets and staff rooms. The shop specialises in the sale of high-quality furniture from a variety of manufacturers. It also has its own corners for specific companies such as Kartell and Herman Miller.

The project is conceived as a large design space in which the architecture allows the viewer to see how the building is constructed. As the shop is of such large dimensions and has to define for itself a proper relationship between investment and yield, it was planned to keep the beams, pillars, fittings and roof exposed. They form a style of architecture that shows the method of construction and also creates a contemporary and technological architectural language, rather than being merely decorative. The choice of construction method, with all elements visible, was not for economic but for aesthetic reasons, in order to create a distinct identity for the shop and reaffirm its brand name.

Great importance was therefore attached to the façade as a means of creating a distinctive appearance – marking it out from the rest of the shopping centre – and to other features such as the recessed area for the shop

商店位于里约热内卢一家家居购物中心内，总面积为970平方米，包含展示厅和会议室、仓库、洗手间、员工休息室等设施。商店专营各种品牌的高品质家具，还设有Kartell、Herman Miller品牌专区。

项目是一个巨大的设计空间，人们可以从其中看到建筑师如何构造的。由于项目规模巨大，如何合理分配投入产出比现代异常重要，最终，建筑的廊柱、配套设施、天花板全部裸露在外。它们形成了一种建筑风格，反映了建筑方式，同时也创造了一种现代技术建筑语言，已经不仅仅是一种装饰。这种建筑方式不是为了节约开支，而是为了美观，通过为商店打造独特的个性来增强品牌的辨识度。

墙壁的设计也凸显了特色，让商店从购物中心中脱颖而出。其他特征——如嵌壁式橱窗店面也十分突出，这一设计打造了一个向商店内部的展示区开放的空间，连接了展示户外产品的独立平台区。这个空间的打造有助于空间的延续性和室内外的过渡，并且保留了项目的设计特色。

橱窗几乎覆盖了商店的全部立面，提高了设计产品的价值，在建筑内部打造了不寻常的特征，也增加了产品和品牌的曝光度。橱窗还连接了室内外，改变了内部视野，并且吸引了消费者的眼球，让他们无论是在店内还是店外都可以注意到摆放在橱窗的产品。橱窗由一整块长条玻璃制成，看似与结构支柱相分离，这种极具现代感的设计展现了建筑结构，创造出一种流畅而界限清晰的建筑语言。

1. The wood reception
2. The whole design of the store is highly modern
3. People could see the goods in the store through the bright display windows
4. The exposed ceiling and columns feature the unique store design
5. Different areas transit naturally and softly
6. The open space provides customers with more comfortable shopping experience
7. The entrance

1.木制的接待台
2.店内整体设计极具现代感
3.明亮的玻璃橱窗让人们在店外也可以注意到店内陈列的产品
4.天花板和廊柱裸露在外形成店内独特的设计风格
5.各个空间连接自然柔和
6.宽敞的展示区让顾客有更舒适的购物体验
7.商店入口

window frontage. This creates a space opening onto an exhibition area inside the shop and forms links with the independent deck area, designed exclusively for outdoor products. The creation of this space favours continuity and the transition from exterior to interior, without altering or removing the character of the project's language.

The creation of a display window covering almost the whole façade of the shop, increases the value of the designer products, creating an unusual feature in the side of the building and drawing attention to the products and brand name, prominently displayed on the building in block capitals. It also connects the interior and the exterior, transforming the internal view and drawing the eye into the large display window where the products are always clearly on show to whoever is inside or outside the shop. The character of the display window is that of a single, long sheet of glass, seemingly disconnected from the structural pillars, the modernist approach releasing and revealing the structure to create a fluid and well-defined architectural language.

1. Deck 1	1.露台1
2. Garden	2.花园
3. Deck 2	3.露台2
4. Store	4.商店
5. Deck 3	5.露台3
6. Lavatory	6.洗手间
7. Kitchen	7.厨房
8. Cloakroom	8.衣帽间
9. Mall	9.林荫道

NOVO AMBIENTE

4

5

6

7

1

Robert Kuo Retail Store

罗伯特·郭零售店

Location:
New York, USA

Designer:
TEK, Charles Thanhauser

Photographer:
Brian Rose

Completion date:
2007

项目地点：
美国 纽约

设计师：
TEK设计公司，查尔斯·丹诺瑟

摄影师：
布莱恩·罗丝

完成时间：
2007

Robert Kuo, a Los Angeles-based furniture designer known for his cloisonné and lacquer work, engaged TEK to design his Manhattan retail store. The client wanted his new east coast store to reflect the feel of his west coast store – clean, simple planes, a minimalist approach to space division, and straightforward material usage – with an identity that is quintessentially New York.

Refined materials, exaggerated scale and axial connections comprise the concept for the design. TEK created an environment at the scale and level of finish of Kuo's furniture and decorative accessories, subtly conveying Chinese traditions through a modern vocabulary. A primary goal for the client was that customers feel at home in the space so that they can picture the merchandise in a residential setting. TEK achieved this by using thick walls and floating, folding walnut ceiling planes to define smaller display areas. The walls never touch the columns and stop short of the ceiling plane in order to express the volume of the space. Base reveals lend a sense of lightness to the walls and provide a clean transition to the floor. Framed views draw visitors from one space to the next and some walls feature niches and openings for displaying special pieces. The most dramatic aspect of the design is the entry sequence which frames the focal point, a 1.5 tons of repoussé copper panel and a water feature, both provided by Robert Kuo, setting the tone for a tranquil, harmonious space.

罗伯特·郭是洛杉矶知名的家居设计师，以设计景泰蓝和漆器作品而闻名。他委托TEK事务所为他设计曼哈顿零售店。他希望自己新开的东海岸店能够延续西海岸店的风格——平面简洁、空间低调简单、材料明确，同时又兼具纽约特色。

精致的材料、夸张的规模和轴向连接共同组成了项目设计。设计师打造了一个与罗伯特·郭所设计的家具和装饰品相符的空间，以现代方式微妙地传达了中国传统文化。罗伯特·郭的主要目的就是打造一种居家感，以便于顾客可以在家居环境中挑选自己喜爱的产品。设计师运用厚厚的墙壁和悬垂的胡桃木天花板界定了小型展示区。墙壁与柱子和天花板是隔开的，展现了空间感。底部打光为墙壁增添了亮度，形成了与地面之间简洁的过渡。框架视图让顾客目不暇接，一些壁龛式墙壁可以展示特殊的作品。设计最夸张的部分是店面入口，由罗伯特·郭所提供的1.5吨的压花铜板和水景装置奠定了空间宁静、和谐的基调。

项目采用了简单的材料——涂漆墙壁和石地面，让焦点集中在罗伯特·郭的作品上。厚厚的墙壁被漆成了温暖的灰白色，明亮、新鲜而又不刻板。设计师通过研究罗伯特·郭的作品而选择了最合适的背景色。性价比极高的石地面可以抵御气候和磨损，其风格让人想起了传统的中国庭院。

2

1. The store is in a clean and calm style
2. The space design combines perfectly with the products
3. The off-white walls divide the space
4. The display of the objects shows a Chinese style
5. This home-style space provides a more relaxing atmosphere for customers to choose his/her favourite product

1.店内风格简洁平静
2.空间整体设计风格与店内产品和谐融合
3.灰白色墙壁用来隔断空间
4.物品陈列具有中国风格特色
5.居家风格能让顾客更放松的挑选自己喜爱的产品

The material concept is a simple palette of painted walls and stone flooring that focuses the spotlight on Robert Kuo's work. The thick walls are painted a warm grey-white, which feels bright and fresh but not stark. Sample colours were viewed in the context of Robert Kuo product colours to determine the most appropriate backdrop. Cost effective stone flooring materials were chosen to hold up to weather and wear conditions as the space is directly entered from the street. The pattern evokes traditional Chinese stone courtyards with a modernist sensibility.

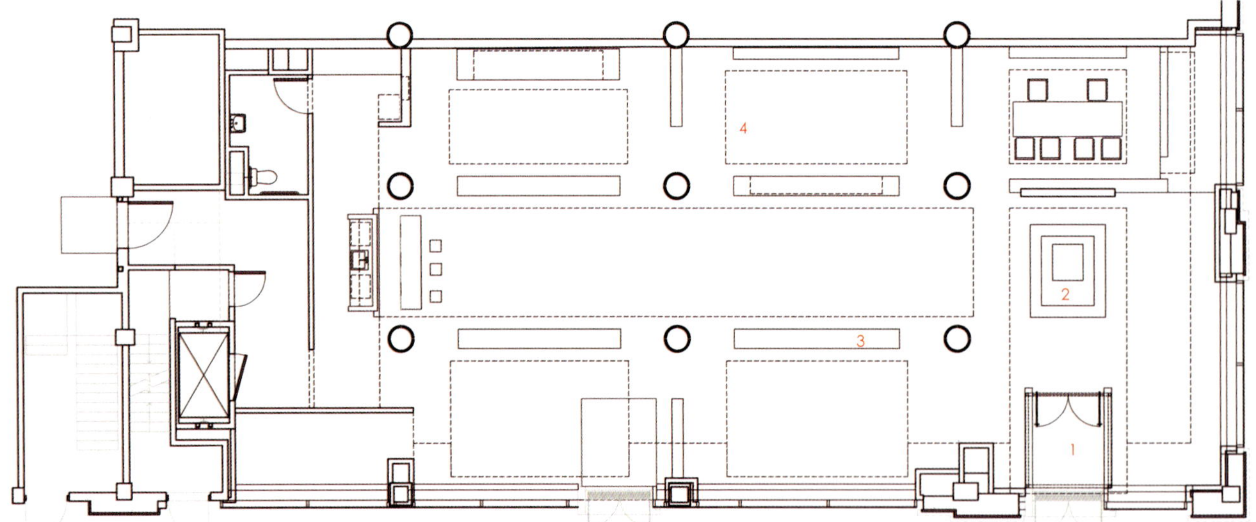

1. Entrance 1.入口
2. Water feature 2.水景
3. Wall 3.隔断
4. Display area 4.展示区

3

4

Showroom Saporiti Italia

意大利萨普利提展示厅

Location:
Shanghai, China

Designer:
Éric Maria Architecte

Photographer:
Thomas Weuthen

Completion date:
2010

项目地点：
中国 上海

设计师：
埃里克·玛丽

摄影师：
托马斯·乌登

完成时间：
2010

Located in prestigious Wending Plaza Design Centre in Puxi, Xuhui District, Si Trading Co. Ltd. is the Saporiti regional base for its activities in the entire region of China. The Saporiti China showroom and contracting office provides a wide range of professional services to the Saporiti Italia Asian clients and to their architects and designers. The showroom displays on a surface of over 600 square metres on three floors the Saporiti Italia best-sellers as well as the innovative Saporiti Kitchens collection. The showroom itself, with its innovative structures and decorative elements, is a design laboratory showing the advanced design capabilities and technological skills of the Saporiti Group.

The showroom explores the multiplicity of Shanghai's atmospheres through a creative fusion of oriental and Western ambiances. A sumptuous urban garden atrium gives the visitor a general overview of the three different levels in the showroom. Each level includes different display rooms arranged by living, resting and working thematic. They are articulated around sophisticated partition walls which are the key elements to create and combine multiple exhibition spaces. Wooden brick walls are inspired in the delicate wooden façade traditional artwork, providing a subtle transition between intimate and public spaces, evoking the unique values present in the Asian culture. These walls guide the visitors through.

Si贸易公司位于徐汇区浦西繁华的文定广场设计中心，是萨普利提在中国的总代理。萨普利提的中国展示厅和办公室为萨普利的亚洲客户以及他们的建筑师和设计师们提供全套的专业服务。展示厅共分三层，总面积600平方米，里面展示着萨普利提的畅销产品和萨普利提创意厨房系列。展示厅通过自身的创意结构和装饰元素展示了萨普利提集团的先进设计理念和工艺。

展示厅通过融合东西方氛围来探索了上海文化的多样性。华丽的城市花园中庭让来访者对展示厅的三层结构有了一个大致的了解。每层楼都有不同主题的产品展示，分别为：生活、休息和工作。它们之间由精致的隔断隔开，这些隔断是打造和混合不同展示空间的重要因素。木砖墙壁的灵感来自于传统艺术品精致的木制表皮，为公共和私密空间提供了微妙的转换点，体现了亚洲文化的独特价值。这些隔断引导着来访者。

生活区的断线设计和红黑相间的大理石地板将来访者带入了一个动感的空间。滑动隔断墙上上海城让空间拥有了熟悉的城市背景。二楼的睡眠区采用了竹子地板，更显温和。悬垂的竹林和红灯笼在更衣室和休息区之间形成了独特的隔断。工作区的图书墙充满了色彩与开合构造的活力。图书馆内一扇无形的转动门通往私人办公区。

2

1. The showroom is set off by various green plants
2. The unique-designed red chair
3. The lighting bands on the ceiling make the space dynamic
4.The quiet winding path in the entrance
5. It is a generous interior space
6. The suspended bamboo forest with Chinese red lantern creates a unique partition wall between dressing rooms and lounge areas
7. The exterior view of the showroom

1.绿色植物掩映中的展厅
2.红色椅子造型个性独特
3.天花板上的条状光带明亮动感
4.入口处蜿蜒幽静的小路
5.室内宽敞开阔
6.悬垂的竹林和红灯笼在更衣室和休息区之间形成了独特的隔断
7.商店的外部造型

In the "living area", the light broken lines and the red and black marble floor arrangement are setting the visitors in a dynamic approach; Shanghai City printed on the sliding partition wall gives a familiar urban background to the space. The "sleeping" area on the second floor is treated with warmer materials as bamboo flooring. The suspended bamboo forest with Chinese red lantern creates an intimacy partition wall with dressing rooms and lounge areas. On the "working" level an extensive library wall is vibrating energy with a play of different colours and open/closed cases arrangement. A surprising invisible pivot door opens in the library an entrance to the private office spaces.

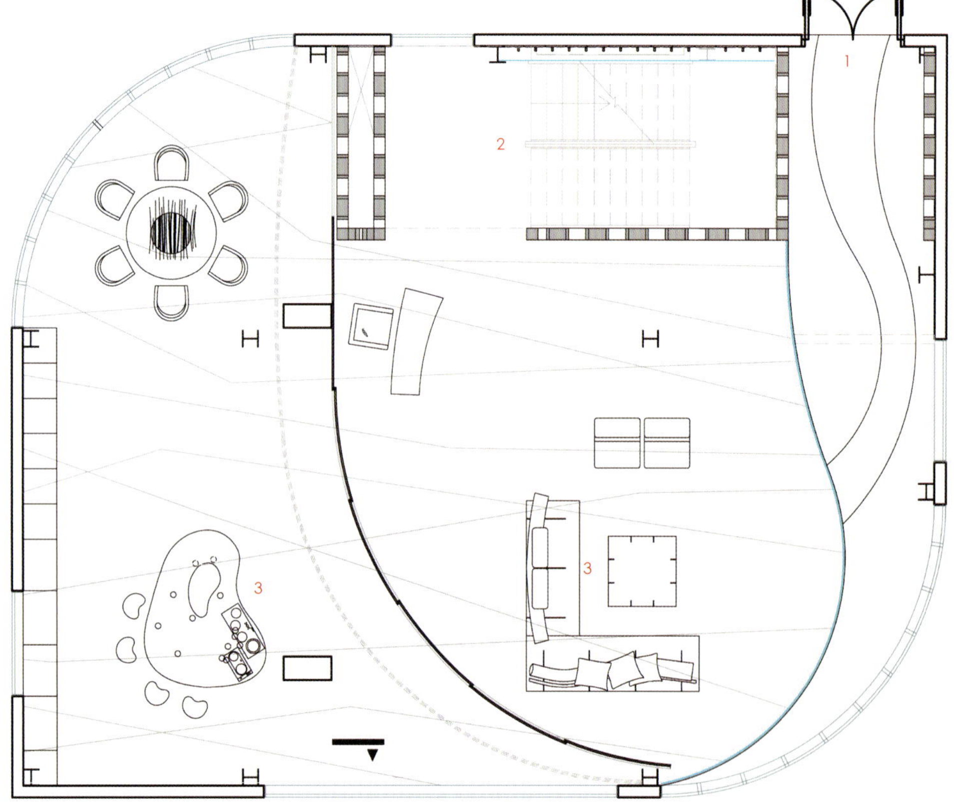

1. Entrance 1.入口
2. Stairs 2.楼梯
3. Showing area 3.展示区

3

4

5

6

7

1

Studio Hagen Flagship Showroom

哈根工坊旗舰店

Location:
Oslo, Norway

Designer:
Ralston & Bau

Photographer:
Studio Hagen

Completion date:
2009

项目地点：
挪威 奥斯陆

设计师：
罗尔斯顿&鲍设计公司

摄影师：
哈根工坊

完成时间：
2009

At the end of August 2009 the 1,000-square-metre space was inaugurated by the mayor of Oslo, Fabian Stang, and accordingly the showroom has been one of the main attractions for the 2,000 visitors of the Designers Saturday in September 2009.

Norway's leading stair producer, Hagen, chose Ralston & Bau to transform an old tramway garage in Skøyen, Oslo to create its flagship showroom. Other producers for doors (Norne) and floors (Solidfloor) were involved to give a complete offer for architects and house producers.

The concept was inspired by the works of the artist M. C. Escher, who used stairs as a major subject in his art. When you enter the showroom you arrive at the black framed entrance of the Stair Garden – before your eyes different stairs are leading up to the second floor. The designers played with the product dimensions and gave each stair a fitting surrounding with individually styled rooms, for example a Japanese inspired interior with a dark laqued stair and a happy family home entrance, where the stair way becomes a playground. Meeting spaces are included among the stairs.

The designers wanted to design a studio that gives you the exciting experience of trying a stair way, walking further and finding a totally different way back. The whole Stair Garden is an invitation to discover the product in a human environment.

2009年8月末，奥斯陆市市长费宾·斯唐亲自为总面积为1,000平方米的哈根工坊旗舰店揭幕，随后，这家店面便成为了2009年9月“设计师周六”活动的主要景点，被2,000余名游客参观。

哈根是挪威的顶级楼梯制造商，他们委托罗尔斯顿&鲍设计公司将一个旧车库改造成新的旗舰店。旗舰店还汇集了诺乃门业和固地地板，为建筑师和房屋制造商提供了全套的服务。

项目的设计理念灵感来自于M.C.艾舍的作品，他使用楼梯作为作品的主要元素。一进展示厅便是楼梯花园的黑色边框入口，里面不同的楼梯引领着顾客上到二楼。每个楼梯都有合适的尺寸，并且配备了各具风格的房间，例如日式房间配备黑木楼梯，其乐融融的大家庭采用可以作为操场的楼梯。楼梯之间是集会空间。

设计师想要打造一个令人难忘的体验工坊，人们通过楼梯上楼，然后通过不同的路径走下来。整个楼梯花园让人们在人性环境中进行发现心仪的产品。

STUDIO
HAGEN
ESCALIA

1. Different stairs leading to the upper floor
2. The black framed entrance of the Stair Garden
3. The space is clean and spacious
4. The special distribution provides customers with more comfortable shopping experience
5. The wide stairs with white armrests look gorgeous and elegant
6. The Stair Garden at the upper floor

1.不同的楼梯展示通向二楼
2.楼梯花园的黑色边框入口
3.整体空间干净宽敞
4.室内空间布局合理，引导顾客更加舒适的体验产品
5.宽阔的楼梯搭配白色的扶手风格高贵典雅
6.二楼的楼梯花园

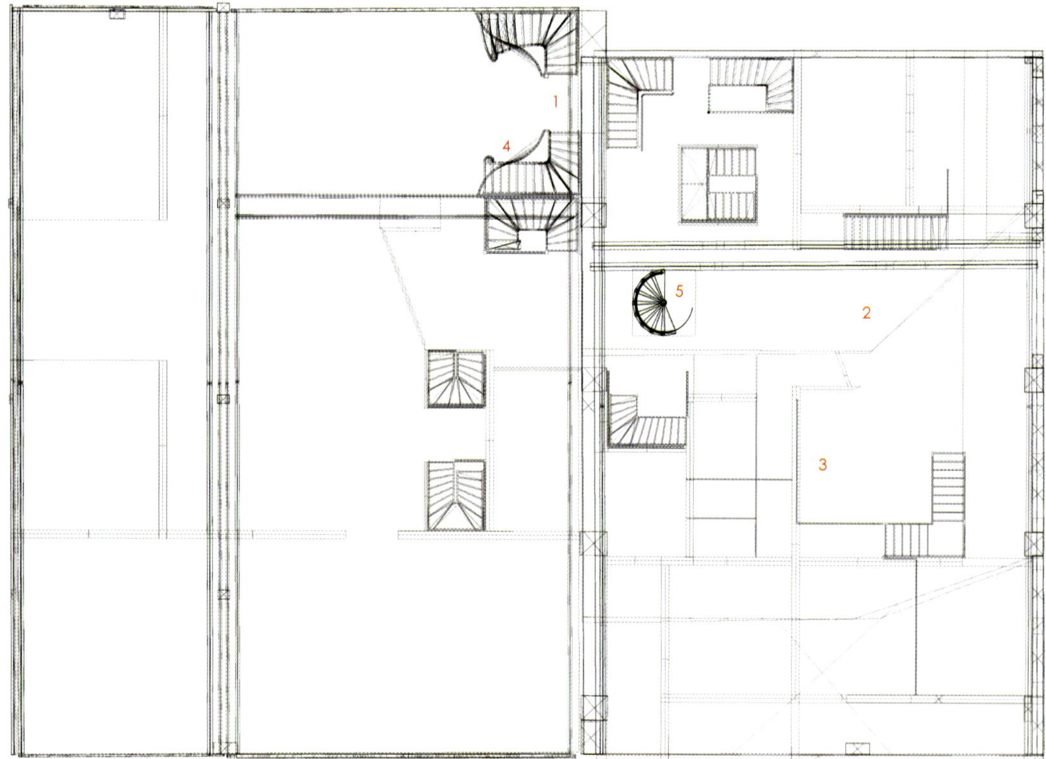

1. Stairs to the upper floor 1.上楼楼梯
2. Entrance 2.入口
3. Reception 3.前台
4. Stairs 4.楼梯
5. Installation 5.装置室

ESCALIA®

4

5

6

1

The Orange Cube

橙色立方

Location:
Lyon, France

Designer:
Jakob + MacFarlane Architects

Photographer:
Nicolas Borel

Completion date:
2010

项目地点：
法国 里昂

设计师：
Jakob + MacFarlane建筑事务所

摄影师：
尼古拉斯·波莱尔

完成时间：
2010

The project was designed as a simple orthogonal "cube" into which a giant hole is carved, responding to necessities of light, air movement and views. This hole creates a void, piercing the building horizontally from the river side inwards and upwards through the roof terrace.

The cube, next to the existing hall (the Salins building, made from three arches) highlights its autonomy. It was designed on a regular framework (29 metres x 33 metres) made of concrete pillars on five levels. A light façade, with seemingly random openings was completed by another façade, pierced with pixilated patterns that accompany the movement of the river. The orange colour refers to lead paint, an industrial colour often used for harbor zones.

In order to create the void, Jakob + MacFarlane Architects worked with a series of volumetric perturbations, linked to the subtraction of three "conic" volumes disposed on three levels: the angle of the façade, the roof and the level of the entry. These perturbations generate spaces and relations between the building, its users, the site and the light supply, inside a common office programme. The first perturbation was based on direct visual relation with the arched structure of the hall, its proximity and its buttress form. It allows to connect the two architectural elements and to create new space on a

项目是一个立方体，上面巨大的孔洞满足了采光、通风和取景需求。孔洞从河面一侧斜插过来，直通屋顶平台。

这座位于大厅（萨兰斯大楼，由三个拱形组成）旁边的立方体建筑，其框架为29米x33米，共分为5层。轻质外墙外面还有一层表皮，上面有河水涟漪的图案。橙色让人想起了一种海港区域经常使用的工业铅丹涂料。

为了打造空隙，设计师运用了一系列的介入体（即孔洞），在三个层面上连接了三个圆锥体结构：外墙、屋顶和入口层面的角度。这些孔洞在楼内形成了空间，联系了建筑、使用者、场地和采光。第一个孔洞与旁边的拱形大厅拥有视觉联系，从造型上联系了两座建筑，并且在建筑内部打造了一个双层空间。另一个是椭圆形孔洞打破了建筑的规律，从河水一侧的建筑一角一直纵向延伸至四层楼。

平台拥有充足的采光和优美的风景，可以从各层的阳台进入。人们可以在平台上进行愉快的交谈，还可以享受美丽的风景。平台的透明度和采光让工作空间看起来更加优雅明亮。最上面的一层是巨大的平台，可以俯瞰里昂、芙丽叶教堂和里昂河的风景。

2

1. The orange building is quite eye-catching
2. There are holes of different sizes on the wall to display products
3. The façade reminds people of ripples in the river
4. The open and spacious interior display area
5. The holes on the façade provide sufficient natural light
6. People can enjoy a beautiful landscape on the terrace

1.橙色的主体建筑十分明亮耀眼
2.墙体上大小不一的孔洞用来展示产品
3.外层表皮让人想起河中漾起的涟漪
4.宽敞开放的室内展示空间
5.外墙孔洞利于室内采光
6.平台上可以欣赏到优美的风景

double height, protected inside the building. A second, obviously an elliptic one, breaks the structural regularity of the pole-girder structure on four levels at the level of the façade corner that gives on the river side.

The tertiary platforms benefit from light and views at different levels with balconies that are accessible from each level. Each platform enjoys a new sort of conviviality through the access on the balconies and its views, creating spaces for encounter and informal exchanges. The research for transparency and optimal light transmission on the platforms contributes to make the working spaces more elegant and light. The last floor has a big terrace in the background from which one can admire the whole panoramic view on Lyon, la Fourvière and Lyon-Confluence.

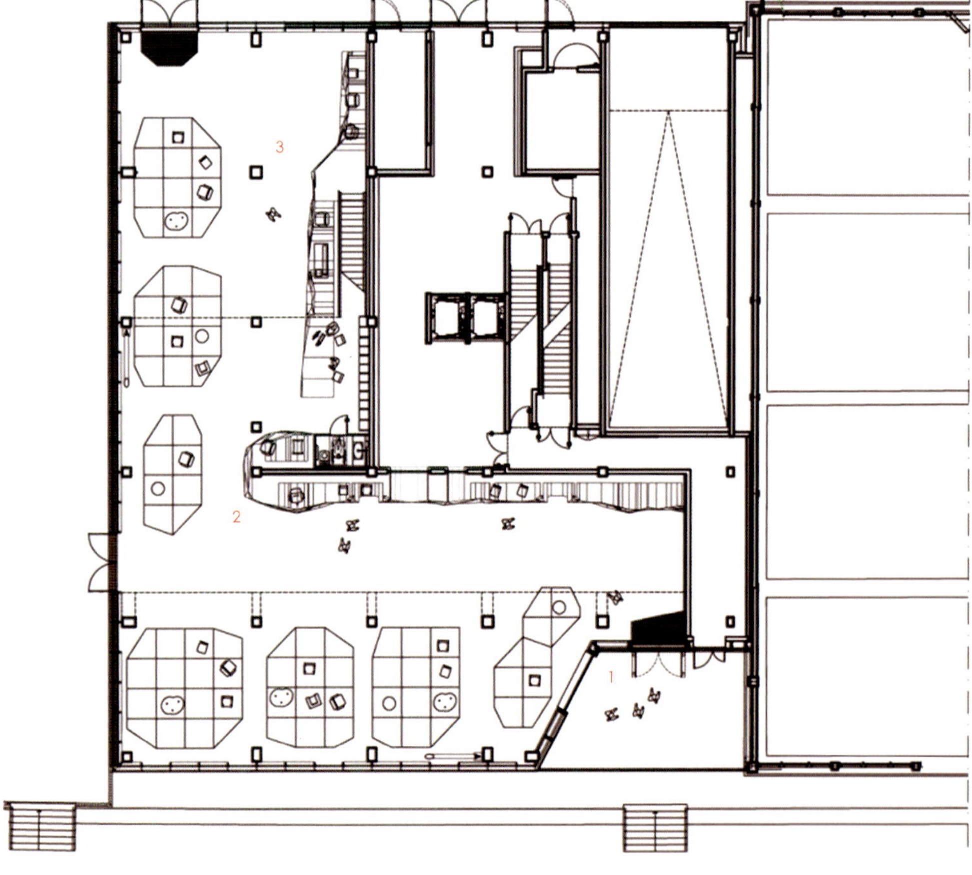

1. Entrance 1.入口
2. Showing area 2.展示区
3. Stairs 3.楼梯

3

4

1

The Reverse Courtyard-Design Republic Flagship Store

设计共和旗舰店

Location:
Shanghai, China

Designer:
Lyndon Neri and Rossana Hu

Photographer:
Derryck Menere and Liu Shenghui

Completion date:
2006

项目地点：
中国 上海

设计师：
郭锡恩，胡如珊

摄影师：
德里克·门奈列，刘圣辉

完成时间：
2006

Design Republic Flagship Store gathers a series of top-design furniture. Located in the historical Bund in Shanghai, occupying 600 square metres, the store is a place of modern classic and pioneering design works. The space includes souvenir shop, private shopping area, exhibition area and collection showroom.

The designers combine the intermittent interior walls and different floors perfectly, creating a unique space. The centre of the building is converted into a courtyard, which is seemingly meaningless but actually solves the problems caused by the building's structure. Adhering quintessence of Chinese architecture, the design uses glass panels and steel frames to change the original structure. This individual backdrop will perfectly set off the modern classic works.

In the middle of the "courtyard" is a wood platform, which is also the most important part of the courtyard, emphasising the main exhibition area. This design makes Design Republic to be a platform of academic communications. Wood structures expand from the entrance to the platform and finally curl up to the ceiling, expressing a sense of dynamic and energy.

In order to show more design works, the designers provide a interactive form for the exhibition area, bringing customers a unique

设计共和旗舰店集合了一系列的顶级设计家居产品。设计共和旗舰店坐落在历史悠久的上海外滩，面积达 600平方米，是现代经典和当代先锋设计的荟萃之地，其空间内还同时容纳了旅游纪念品、私人购物区、展览区以及收藏品陈列室。

设计师将整个建筑内部断续的墙壁同错落的楼层完美结合，勾勒出了一个独一无二的完美空间。建筑的中间部位被改建成一个庭院，这看似无心的设计解决了建筑本身结构引发的问题。设计秉承了中华建筑的精髓，并通过玻璃面板以及钢制框架改变原有结构。这一独立的背景，将使那些现代经典得到完美的呈现。

在“庭院”的正中，或者说整个“庭院”中最重要的地方，有一个木质结构的中心平台来强调重点展区，从而烘托整个设计共和使之成为一个学术性的设计交流平台。木质结构从入口处一直延伸到平台，然后卷起连接到天花板，整体看起来动态十足，动力源源不绝。

为了展示更多的设计，设计师将展区设计成了互动的形式，给顾客带来一种与众不同的购物体验。购物区、长廊以及各个展区是分别与中心区域相连的独立个体，这种设计给人带来了独特的经历和感受。

2

1. The wood floor and ceiling echo each other
2. The display area in grids
3. The open and bright space provides customers with a great space for shopping
4. Walls in glass panels and steel frames
5. The flexible display of products adds energy and dynamic into the space
6. The courtyard display platform in the centre of the store
7. Products in the display windows

1.木质地板与天花板互相呼应
2.栅格状的展示区
3.空间宽敞明亮，给顾客充分的体验空间
4.钢制结构与玻璃面板形成的断续墙壁
5.灵活摆放的产品让空间充满动感与活力
6.店内中央设计的庭院式展台
7.橱窗里的产品展示

shopping experience. Shopping area, gallary and various exhibition areas are all individual elements connected to the centre. This design provides customers with distinctive experience and feelings.

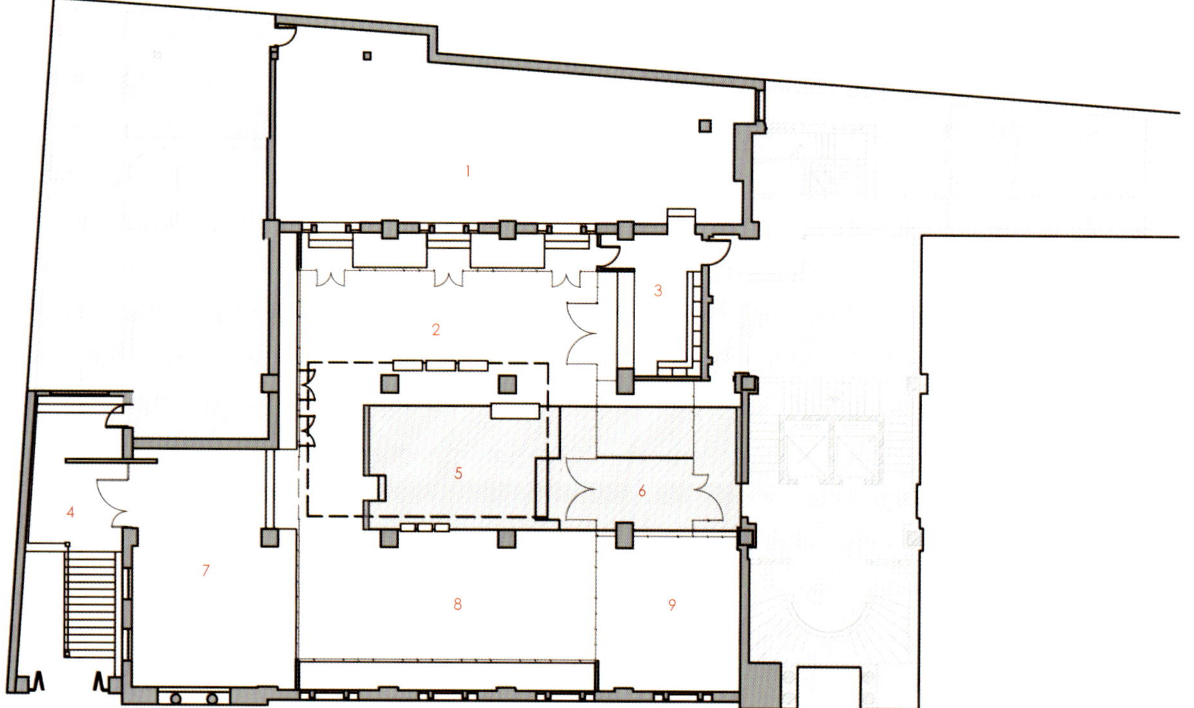

1. High-end merchandise 1.高端商品
2. Classic 2.经典商品
3. Reception 3.前台
4. West entry 4.西入口
5. Exhibition area 5.展示区
6. Main entry 6.主入口
7. Objects 7.商品
8. Classic 8.经典商品
9. Traveller's objects 9.旅行者商品

4

5

6
7

1

VilaSofa

维拉沙发专卖店

Location:
The Netherlands

Designer:
Tjep.

Photographer:
Frank Tjepkema

Completion date:
2008

项目地点：
荷兰

设计师：
Tjep设计公司

摄影师：
弗兰克·杰珀柯玛

完成时间：
2008

Tjep. was commissioned to create a shop environment for a new furniture brand called VilaSofa. VilaSofa is positioned between a conventional furniture shop and IKEA. It offers mid-range prices for a wide public and a 48-hour delivery guarantee for all displayed models. The 48-hour claim became the main theme for the shop. Designers used the idea of a warehouse as metaphor for speed: a place where goods are stored for the transition from the producer to the very personal environment of your home.

The conceptual solution was surprisingly simple and logic: a combination of warehouse esthetics and home esthetics. This resulted for example into materials such as plywood usually used for crates combined with high-end glossy finishes. The symbols used in transportation and packaging have been metamorphosed into decorative elements that form room dividers that were arranged to organise the space and routing. Finally there is a big wall suggesting the idea of a magical VilaSofa. This is the only place where the graphic references are more literal: a big chandelier, a diversity of playfully arranged windows, romantic balconies. In this space designers placed big "picnic" tables where clients can take their time to inform about the product, talk to the VilaSofa staff. Finally designers designed mobile cashing

Tjep设计公司接受了一个新的家具品牌“维拉沙发”的委托，为其装修店面。这一品牌的定位介于传统家具店和宜家之间。商品为平宜近人的中档价位，并承诺所有商品48小时内送货上门。这是店铺的宣传重点。设计师以仓库的理念来比喻送货迅速：这里只是厂家的存货地点，将会直接送到您的家中。

设计构想很简单：将仓库与家庭结合起来。材料采用木制包装常用的胶合板，配以高档的光泽感装饰。运输与包装在这里演变成装饰元素，划分并合理安排了空间。有一面巨大的墙壁暗示着专卖店的童话色彩，墙上刻画着具体的图形：大吊灯，随意排列的窗户，还有浪漫的阳台。设计师在这里放了一张巨大的野餐桌，顾客可以在这里与店内工作人员探讨商品。设计师还设计了移动付款方式，顾客可以坐在新买的沙发上付款。店内不仅有大众款式的沙发，还有莫尼卡·穆尔德、柯迪·菲斯等著名设计师专门设计的款式。

2

1. There are rich decorations in the store
2. The fairy-tale white wall
3. The window is decorated with a circle of bulbs
4. The products are cosy and comfortable
5. It is a place full of romantic elements

1.室内丰富的装饰元素
2.白色的墙壁充满童话色彩
3.圆形的彩灯装饰的窗子
4.产品风格平易近人，随意舒适
5.空间装饰浪漫感十足

systems so that clients can actually pay from their newly adopted sofa. VilaSofa offers a combination of existing sofa models and specially designed sofas by designers such as Monica Mulder and Khodi Feiz.

1. Display area 1.展示区
2. Main hall 2.主大厅
3. Stairs 3.楼梯
4. Sofa stand 4.沙发台
5. Rest area 5.休息区

3

4

1

2006Feb01

2006.02.01概念店

Location:
Vienna, Austria

Designer:
BEHF Architects

Photographer:
Bruno Klomfar

Completion date:
2006

项目地点：
奥地利 维也纳

设计师：
BEHF建筑事务所

摄影师：
布鲁诺·克洛姆法

完成时间：
2006

BEHF came up with a design that takes into account the attractive and unique location in the heart of a lively and busy shopping metropolis as well as the historical building fabric with its courtyards and old, elegant rooms. Although the store has views to St. Stephen's Cathedral and the baroque Donner fountain in the Neue Markt, and is located at the intersection of Kärntner Straße and Graben, Vienna's most important shopping streets in the centre of the old city, the theme of the two-storey store is guided by a historical Capuchin monastery garden. These two worlds – the busy streets outside and the stillness of the garden – are the framework for the store's "living area" in that both the street zone and the garden courtyard are woven into the structure and design of the store.

The historical stone façade of the former bank building was playfully re-interpreted for its new purpose, with elegant portal elements painted black and transparent, semi-transparent mirrored glass sections that can be opened. The idea of transparency and lightness continues inside the store by means of movable organza curtains so that the hidden green of the garden can shimmer through; at the same time permitting an atmosphere of "privacy" and "discretion".

The hardware – floors, stairs, walls and ceilings – will bear the typical BEHF signature: untreated,

项目位于繁华而生机勃勃的商业中心，地点独特而具有吸引力。店铺坐落在一座古老的建筑之中，带有庭院和优雅的房间。从店内可以看到圣史蒂芬大教堂和巴洛克风格的唐纳喷泉，并且地处维也纳两条最繁华的商业街的中心。两层高的店铺被设计成古老的修道院花园风格。繁华的街道和宁静的花园是两个世界，二者共同形成了店铺的界限，融入了店铺的结构和设计之中。

建筑原来是一座银行，设计师重新诠释了店铺的门面，用黑色和透明玻璃装饰了优雅的门口，半透明的镜面玻璃可以任意开合。可移动的透明硬纱帘让店内设计延续了透明元素和轻盈感，既可以隐约看见花园的绿植，又可以保证私密的气氛。

地面、楼梯、墙壁、天花板等设计硬件体现了典型的BEHF风格，是未经处理的灰色平滑水泥面。简洁的长方形楼面两侧是由豪华蛇皮包裹的服装间，里面有时刻更新的惊喜元素——壁龛、衣架、抽屉和试衣间都装饰着奢华的材料，如抛光不锈钢和华丽的织锦挂毯。像木偶剧院和舞台一样，这些元素可以随意改变，形成了独特的气氛和专属氛围。

大片的沙发都包裹在织物之中，便于顾客坐下休息。它们被随意地放置在一起，上方是雅观的意大利定制手工吊灯。吊灯将室内灯光调节的异常柔和，适合店内的各种活动。

2

1. The store is located in an old building
2. It has an elegant colour palette
3. Curtains, chandeliers and sofas look gorgeous and luxury
4. The green wall decorations inside the changing room
5. The open space provides customers with enough space for shopping

1.商店位于古老的建筑之内
2.整体色调十分优雅
3.帷幔、吊灯和沙发搭配得高贵奢华
4.试衣间内部的绿色墙饰
5.宽敞的室内让顾客有足够的活动空间

smooth cement surfaces in a purist grey. Two-storey-high suitcase cubes covered in sumptuous, glove-quality reptile leather flank the simple, rectangular space of both floors. Inside they hold ever new and surprising objects. Niches, shelves, drawers and changing rooms are decorated using different luxurious materials, for example polished stainless steel or magnificent tapestry fabric. These can be varied and changed like in a puppet theatre or stage. They create a unique mood and exclusive atmosphere.

Large presentation surfaces are covered in fabric, inviting the customers to sit and relax. They were arranged casually in private groups under tasteful chandeliers handmade in Italy specially for the store. They allow the light to be adjusted discreetly in harmony with the time of day and the activities in the store.

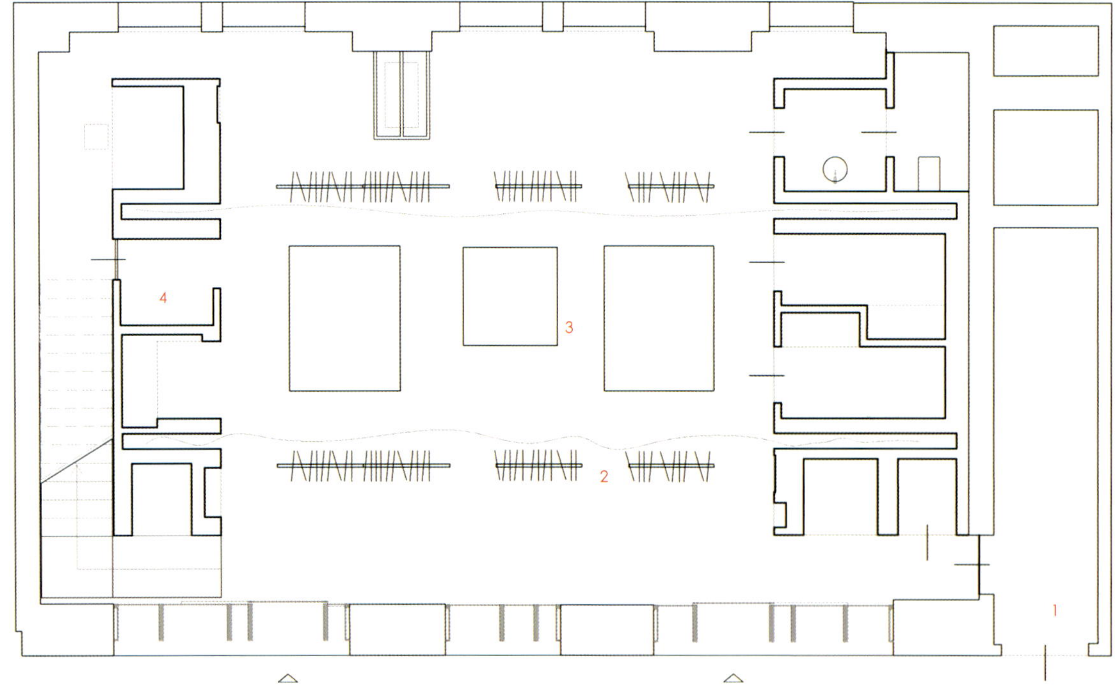

1. Entrance 1.入口
2. Clothes-hanger 2.衣架
3. Sofa 3.沙发
4. Fitting room 4.试衣间

3

4

5

1

Ann Demeulemeester Shop
安·德穆拉米斯特商店

Location:
Seoul, Korea

Designer:
Mass Studies

Photographer:
Yong-Kwan Kim

Completion date:
2007

项目地点：
韩国 首尔

设计师：
Mass工作室

摄影师：
金坤万

完成时间：
2007

The site is located in an alley, at a block's distance from Dosandae-ro – a busy thoroughfare in Seoul's Gangnam District – in close proximity to Dosan Park. Primarily residential in the past, the neighbourhood is undergoing a rapid transformation into an upscale commercial district full of shops and restaurants. The building is comprised of one subterranean level and three floors above. The Ann Demeulemeester Shop is located on the ground floor, with a restaurant above and a Multi-Shop in the basement.

This proposal is an attempt to incorporate as much nature as possible into the building within the constraints of a low-elevation, high-density urban environment of limited space. The building defines its relationship between natural/artificial and interior/exterior as an amalgamation, rather than a confrontation.

The parking lot/courtyard is at the centre of the site, exposed to the street on the southern end. The entrance to the Ann Demeulemeester Shop is located on the western side of the courtyard, and stairs that lead to the other two programmes are located on the eastern side. Landscaping of dense bamboo forms a wall along each of the remaining three sides that border neighbouring sites. Inside the ground floor shop, undulating dark brown exposed concrete forms an organically shaped ceiling. Round columns on the edges of the space continue the ceiling surface while providing

项目位于一个小巷内，距离首尔江南区的繁华大街——Dosandae-ro一个街区，紧邻都山公园。附近街区原来是一个住宅区，现在正逐步被改造成一个店铺林立的高级商业区。建筑为地下一层，地上三层，安·德穆拉米斯特商店位于一楼，上面是餐厅，下面是百货店。

项目力求在各种限制条件下（低立面和高密度城市环境所限制的空间），在建筑中加入最多的自然元素。建筑以融合的方式连接了自然与人工、室内与室外，而不是让它们进行对抗。

场地中间是停车场，南面朝街。安·德穆拉米斯特商店的入口位于庭院的西侧，通往另两个区域的楼梯设在东侧。茂密的竹子景观形成了墙壁，包裹了其他三侧的墙壁。一楼的服装店里，波浪形的深棕色混凝土形成了有机造型的天花板。空间边缘的圆柱延伸了天花板的结构，也提供了必要的结构支撑。这种结构系统打造了不同尺寸的拱形门窗，朝向外面的街道和竹篱。这种有机造型不仅营造了动感的空间，也让空间得以灵活运用。东侧的翼楼里设置着辅助设施，如试衣间、仓库、洗手间，在有效地划分空间同时，又将空间连接了起来。

1. The large glass display windows
2. The building is wrapped with green plants
3. The off-white stairs leading to the upper floor
4. The white display area is impressive in the dark background
5. People can have a view of the street through the windows along the staircase
6. A general view of the sale area

1.商店的大玻璃橱窗
2.建筑外表被绿色植物所覆盖
3.通往楼上的灰白色楼梯
4.白色的展示区在暗色调的背景下十分醒目
5.楼梯两侧的玻璃窗可以看到街道
6.销售区纵览

the necessary structural support. This structural system creates arched openings of varying sizes that are open and as exposed as possible to the outside road and the bamboo hedges. This organic formation is not only a dynamic space but a flexible rectangular one (11.2 metres x 14 metres). The additional wing on the eastern side contains support functions such as fitting rooms, storage and a bathroom, efficiently divided and connected at the same time.

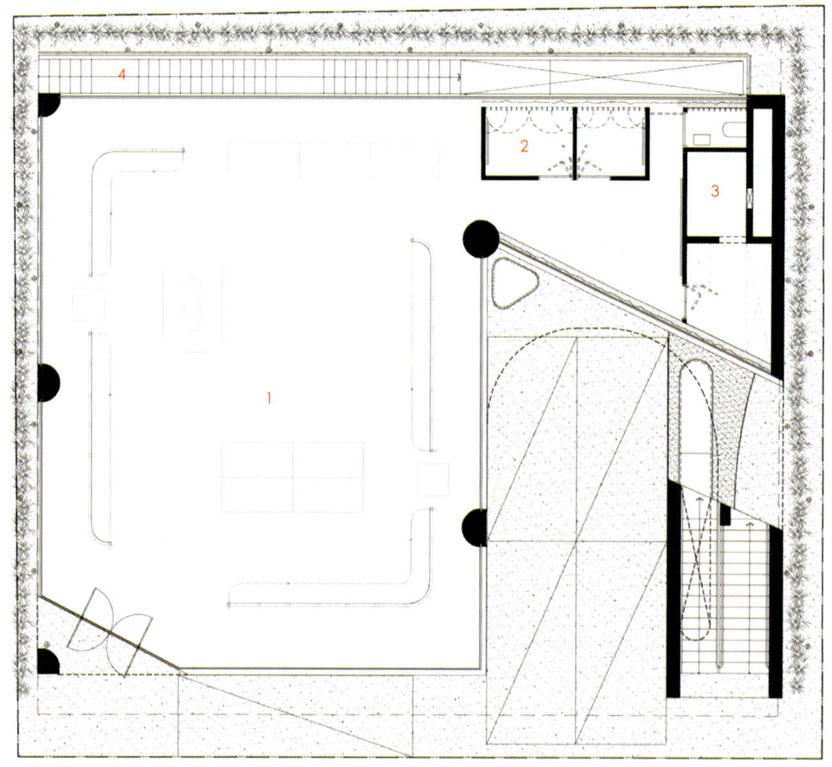

1. Display area
2. Fitting rooms
3. Storage
4. Emergency escape

1.展示区
2.试衣间
3.仓库
4.紧急逃生楼梯

3

4

5

6

1

Aquagirl Shop
水女孩服装店

Location:
Tokyo, Japan

Designer:
Ilaria Marelli Studio

Photographer:
Nacasa & Partners Inc.

Completion date:
2007

项目地点：
日本 东京

设计师：
伊拉莉亚·马瑞利设计事务所

摄影师：
Nacasa摄影公司

完成时间：
2007

The shop sells clothes, shoes and accessories, but the brief tells to give more importance to the shoes, so designers separated three areas and put the shoes area in the centre, on stage.

The shop has to be very feminine and sophisticate, so designers work on bright colours and sort of natural light to make people feel good in a comfortable atmosphere. The heart of the project was to create a surprising environment, thanks to the mirroring effects of the four big swiveling mirrors and thanks to the light effect, which seems to come from inside of the ceiling or the walls and not from real lighting apparels.

The shop has to be perceived as a whole airy space, so designers made some tests about colours of floors, walls and ceiling to match them perfectly despite of the different light on each. It means that floor, walls, and ceiling are all different nuances of light cool grey, but they look exactly the same. The light was very important, since the designers decide not to use any spot light but only diffused light. So they used a fluorescent linear lighting hidden in the border of every false ceiling, and square light with compact fluorescent, inside the false ceiling area. The glossy finishing of the floor helps to diffuse the lighting and the mirrors do the same. Also the big niches have indirect lights hidden in the case and in the top, so to

这间时尚店将出售服装、鞋和配饰，鞋是这里的主要商品，因此设计师划分了三个区域，将鞋子的柜台放在最中央。

店铺的风格需要非常女性化和个性化。设计师利用亮丽的色彩和自然光营造出令人心情愉悦的舒适氛围。 设计的重点是打造一个令人惊艳的店面，设计师安装了四面大镜子，并利用灯光效果，看上去灯光像是从天花板和墙壁中照射出来的一样。

店铺的采光非常好，设计师们为了使地板、墙面和天花板的颜色与不同角度的光照完美地搭配起来，做了很多试验。最后，地板、墙壁和天花板采用了深浅略有不同的浅灰色，但它们看上去却是完全一样的。灯光十分重要，因此设计师决定不使用射灯，而只用漫射灯。设计师在吊顶天花板的边缘安装了灯带，并在吊顶内侧紧挨灯带的位置安装了方形灯。地板光亮的表面也有助于分散灯光，与镜面的效果相同。壁龛里同样装着间接光源，看上去就像墙上嵌着一个发光的长方形。

在设计中，所有的家具都有商业用途。橱柜式长椅和收银台都被漆成亮白色，所有的衣架（包括站立式的和嵌入墙内的）都是不锈钢材质，上有香槟色的涂层。帽子架是雕塑的造型，同样是香槟色的不锈钢材质，底座装有荧光灯带。

2

1. The custom-designed lighting fixtures cast bright and soft lighting for the interior
2. The pure and elegant white colour matches with the products well
3. The bright and clean floor can reflect lighting, improving the brightness of the space
4. The mirrors along the walls will enlarge the space visually
5. A corner of display area

1.天花板上精心设计的灯带使室内光线明亮柔和
2.白色色调纯洁高雅，十分适合店内产品的风格
3.光洁的地面有助于反射灯光，提高室内亮度
4.店内安放了四面镜子，可以扩展视觉空间
5.商品展示一角

be perceived as bright rectangle in the wall.

All the furnitures of the commercial area are on the design. The cabinet benches and casher are in lacquered shiny white, while all the hanger (self standing and fixed to the wall) are in stainless steel champagne finishing. The hat hanger, conceived as a sculptural piece was in stainless steel champagne finishing, with fluorescent light in the base.

1. Entrance	1.入口
2. Mirror	2.镜子
3. Counter	3.柜台
4. Shoes & bag	4.鞋包区
5. Carpet	5.地毯
6. Fitting room	6.试衣间
7. Storage	7.仓库
8. Office	8.办公室
9. Kitchen	9.厨房

4

1

BMW Lifestyle

宝马生活时尚店

Location:
Beijing, China

Designer:
Eightsixthree Architecture Interiors

Photographer:
Elion Yau Ying Ching (Eightsixthree)

Completion date:
2008

项目地点：
中国 北京

设计师：
八六三建筑设计事务所

摄影师：
游英正（八六三建筑设计事务所）

完成时间：
2008

The concept started by taking the sculptural qualities of the favourite recent BMW sports cars, the M1, Z8 & Z4 and to translate the forms of these cars subtly into an innovative retail environment. It is aimed to create a store that had a strong visually dynamic, elegant and timeless shop front and is inspired by the kidney -shaped front grill of BMW cars. The designers wanted to create a shop front that had the elegance and timelessness of this grill design yet create a design that would make one cross the street to see it closer. It was important to also create a large see through area into the store that would allow the clients VM (Visual Merchandising) team a large staging area at the front of the store to create dynamic window displays, as such they crated slots in the ceiling areas and movable platforms in the base of these window areas to aid with their displays.

As with choosing the colour of a new car, the designers went through many options and wanted a classic neutral colour that would allow the colour of products to be shown off to their best. They also felt the colours had to be in tune with current colours used on BMW models and also had to reflect the corporate identify of the client. As such they opted to follow the colours of the BMW motor sport division albeit keeping the colour palette to just two of the three colours reinweiss (white) and signalblau (blue). The diagonal blue stripe

项目的设计概念，是由抽取深受欢迎的最新宝马跑车M1、Z4和Z8 那如雕塑般的外形开始。 然后把这些车的造型巧妙地演绎成一个具有创意的零售空间，目的是来创造一家在视觉上有很强动感，精致和恒久的店，而它的灵感是来自宝马跑车前的腰形进气口， 能让人从街上远处走近来看她。另外重要的是，同时创造一个很大而能透视到店内的空间，让客户的产品展示团队，于店面创造有动感的橱窗展示。如那些于天花的造型灯光凹槽和活动的平台，可帮助他们的展示。

对于店的颜色设计考虑，就如为新车挑选颜色一样，设计师试了很多方案。设计师想有一种能让产品颜色展示它们最好一面的经典中性色调。同时，设计师认为所用的色调，须与现有宝马车系所用的颜色互相调和，并能反映出客户的品牌身份。就此，设计师选用了宝马跑车部门的颜色，并尽量在三种颜色中保留其中两种——白色和蓝色。那于墙面走上天花的蓝色造型斜条，让人立刻认出那是直接演绎了宝马M系跑车的商标。还有， 所有用于店的油漆，是跟车厂使用的一样。

为了当顾客走进店内时，能容易地看见整家店，和货品能清楚地用多重层次来展示，设计师创造了可让产品在高、中及低的高度展示的展架。设计师的构思是让产品能展示于地上，于低处的展柜，于中间的挂架上，或于高处的造型展墙。

所有的展台和附件配置， 是由八六三根据它们的个别用途要求特别设计，包括展台、展柜、碳纤维挂架和造型展墙(原本设计是有156个抽屉或展示层架)。矮展台设计成像时装表演的天桥，来展示行李箱和自行车。那倾斜的展柜是根据宝马M系的商标图案来设计，它可让细小的产品展示于玻璃展柜中。

2

1. The whole store is dynamic, full of streamlines
2. The store is spacious and bright
3. The bright and clean ceiling and floor can reflect lighting
4. The interior design applies BMW's classic colours – white and blue
5. It is an open and bright space

1.整个店面充满着流线型的动感
2.店内宽敞明亮
3.光洁的天花板和地面可以反射灯光
4.室内采用了宝马品牌的经典颜色——白色和蓝色
5.流线型的室内十分通透

running up the wall is an instantly recognisable and direct interpretation of the inclined logo used on BMW M-series cars. All paint used within the store is the same as that used in the motor industry.

Upon entering the store, customers can view from the front to the back of the store easily and to clearly display merchandise at multiple levels. Therefore the designers created furniture that allowed to display products in a low, medium and high level format. The idea was to have products displayed on the floor, at low level in show cases, at mid level hanging from rails or at high level displayed on the feature wall.

All display fixtures and fittings, i.e. podiums, show cases, carbon fibre hanging rails and the feature wall (originally with 156 pull out display shelves) were custom designed by Eightsixthree and are all specific to their individual requirement, low level podiums designed to look like fashion show catwalks are used to display luggage and bicycles, inclined (M-series) display cases are based on the BMW roundel logo and allow display of small items within a glass display case.

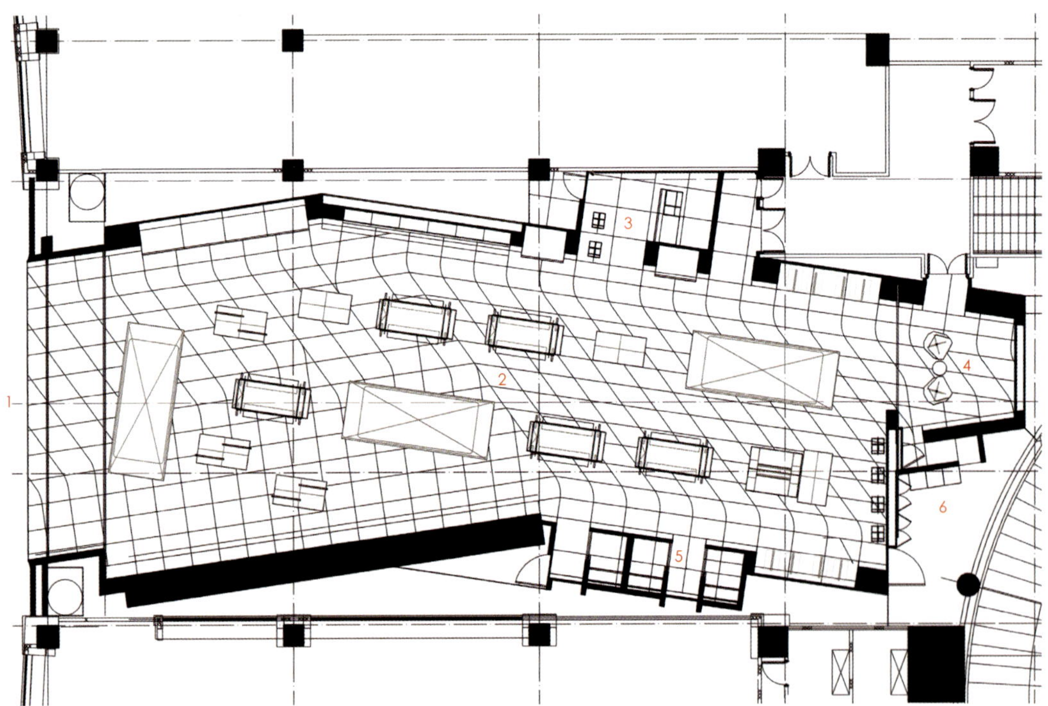

1. Shopfront 1.店面
2. Display areas 2.展示区
3. Cash wrap 3.收银台/包装区
4. VIP lounge 4.VIP休息室
5. Fitting rooms 5.试衣间
6. BOH 6.工作区

4

1

Brother+Sister

兄妹概念店

Location:
Hong Kong, China

Designer:
MARKZEFF

Photographer:
Olaf Mueller

Completion date:
2010

项目地点：
中国 香港

设计师：
MARKZEFF

摄影师：
奥拉夫·穆勒

完成时间：
2010

Brother & Sister is a high-energy concept store located at 1881 Heritage, Hong Kong and designed for Dragon-i founder, Gilbert Yeung and his sister, the Director of Cindy Yeung Emperor Group. The new concept store has a café and bar that offers a modern and highly stylised retail selection of jewellery, watches, fashion, home accessories and exclusive crossover sneakers and limited edition sneakers by top designers and features underground, local brands and young designers, as well as more established and international designers.

Designed to be a visually beautiful store with high gloss charcoal concrete floors and bright white lacquered panelled walls. The furniture is Victorian in style made from distressed black leather and charcoal velvet with contrast red stitching and welting. All of the merchandise is showcased in illuminated vitrines using the same materials as well as accented in burnished bronze. The store is a bit rock and roll with a fresh approach.

兄妹是一家高端概念店，位于香港1881，专门为龙的传人创始人杨其龙和其妹英皇集团总监杨诺思而设计。新概念店设有咖啡厅和酒吧，提供现代而时尚的零售服务，如珠宝、手表、时装、家居饰品和独家的跨界设计的运动鞋、限量版运动鞋等，囊括了顶级设计师、新锐设计师、本土品牌、年轻设计师以及国际知名设计师的作品。

高光泽度的碳化水泥地板和光亮的白色喷漆墙面让店铺内部流光溢彩。家具采用了维多利亚风格，黑色皮革和天鹅绒与针脚和滚边形成了鲜明对比。所有的商品都陈列在同样材质的玻璃橱窗里，下面有光洁的黄铜衬托。店铺融入了一些新式摇滚风格。

2

1. The symmetrical design has a classicism style
2. The products are fashionable and modern
3. The sneaker area
4. When tired, customers can have a rest at the bar
5. The main bar is elegant with flowing light and colours
6. The showcase with unique products

1.室内对称设计具有古典主义风格
2.店铺商品时尚现代
3.运动鞋展示区
4.客人购物之余可到吧台休息小坐
5.主吧台设计高雅非凡，流光溢彩
6.玻璃橱窗展示特色商品

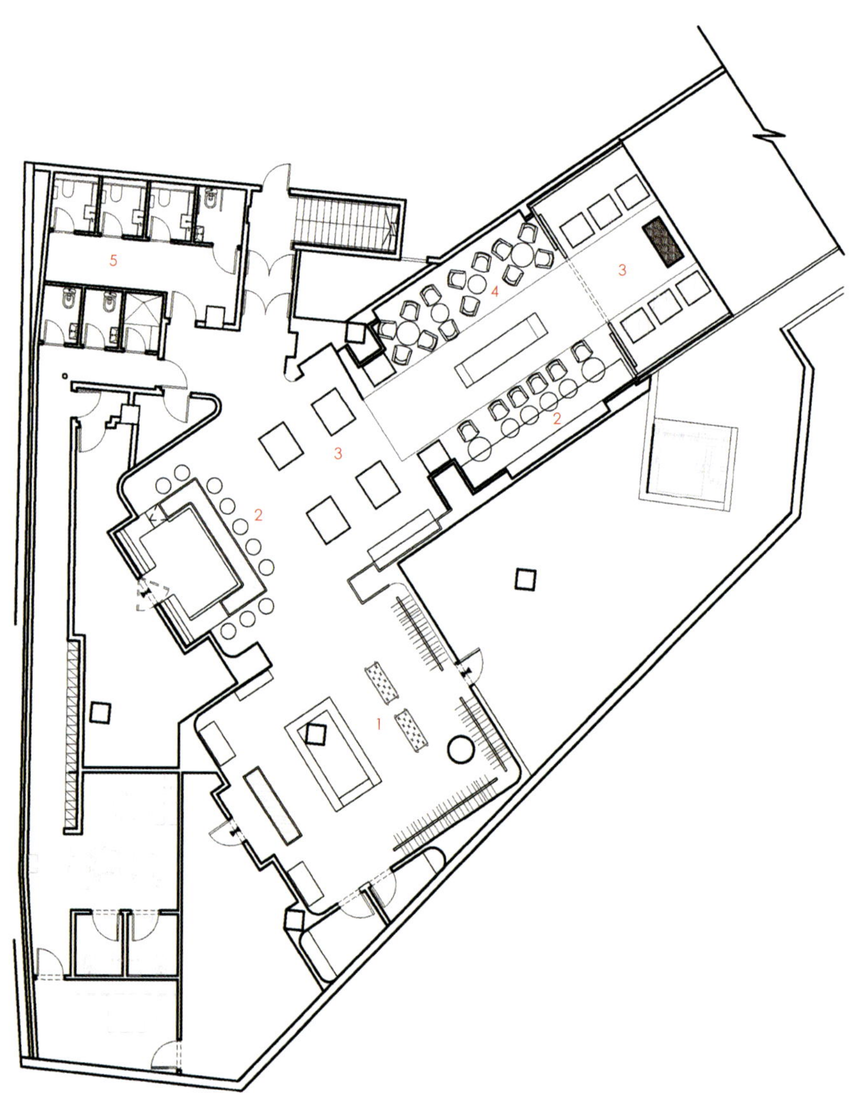

1. Exhibition area 1.展览区
2. Bar 2.吧台
3. Display area 3.展示区
4. Restaurant 4.餐厅
5. Toilet 5.洗手间

3

4

5

6

1

Carlino Agency

卡里诺代理

Location:
Naples, Italy

Designer:
Ilaria Marelli Studio

Photographer:
Communication in Pix+Txt

Completion date:
2009

项目地点：
意大利 那不勒斯

设计师：
伊拉莉亚·马瑞利设计事务所

摄影师：
传播图片文字公司

完成时间：
2009

Italian architect Ilaria Marelli designed the exposition format for the new brand Carlino Agency, a new format born from an idea of the Carlino family, which focuses on the wholesale of the brand RB di Roccobarocco, Krizia Poi, Mario Valentino, Martissima di Marta Marzotto, regarding fashion accessories like bags, suitcases, belts and ties.

The project is situated in a vast area of prefabricated industrial buildings, covering a surface of 1,400 square metres on two levels. "Out of context" is the concept of the project developed by Ilaria Marelli. Natural elements suggest a different dimension and mood of the space than the one given by the standardised industrial building itself. The elements of strongest impact are five "boats" on true scale, reduced to their structural skeleton in walnut. Moored on a resin floor with a natural stone effect, they create the core of the five brands' staging.

Another reference to the "natural-artificial" concept was given by real trunks of three metres height reflecting in full height mirrors, creating an effect of a deep forest. The "tree+mirror" solution can be found both in the shop windows, as a contrast to the industrial landscape outside, and inside the showroom, where they were used in wall niches as impacting display stands for belts and visually enlarging the space.

意大利建筑师伊拉莉亚·马瑞利为新品牌卡里诺代理设计了展览模式，这个新模式源自卡里诺家族。卡里诺家族所经营的品牌包括RB di Roccobarocco, Krizia Poi, Mario Valentino, Martissima di Marta Marzotto，主营时尚配件，如包袋、手提箱、皮带、领带等。

项目位于一座开阔的预制式设计工业楼里，占据两层，总面积为1,400平方米。"脱离周边环境"是项目的设计理念。自然元素营造出一种异次元的空间感，在标准的工业楼十分违和。影响最强烈的元素是五艘真实大小的胡桃木骨架"船"。船停泊在具有天然石材效果的树脂地板上，为五个品牌提供了展示核心。

另一个"自然–人造"概念由真实的3米高树干和落地镜子营造，让人仿佛置身森林深处。"树木+镜子"的方式经常出现在橱窗设计中，与外部的工业景观形成鲜明的对比。在展示厅内部，它们被设计在壁龛里，可以作为皮带的展架，也在视觉上扩大了空间。

除了这些风景和自然元素、装饰，整个空间充满了光洁的白色表面。喷漆货架、箱子和平台的表面营造出一种有趣的反射，表现出一种纯净感和光亮感。灯光的设计模仿了自然光，天花板上的灯光、内置照明灯壁龛和二楼的天窗让空间异常光亮。

带有门厅的一楼提供各种服务：保真验货、付款台和吧台，顾客可以在此获得信息，前往想去的区域。这些服务区与展览区通过胡桃木条墙隔开，营造出一种半透明的效果，既保证了隐私，又不是完全封闭。这些入口的墙壁将人们的注意力引到空间的中心，让顾客通过一个景观楼梯到达二楼的五个展览室里，五个展览室之间也是由胡桃木条墙隔开的。

VALENTINO

1. The white wall is a perfect backdrop for the various products
2. The relaxing area is surrounded by trunks, creating a sense of nature
3. A display shelf like a spiral staircase
4. Wood bar screens divide different areas
5. The walnut skeleton ship has a unique feature
6. The whole place is clean and bright

1.白色墙壁突出映衬了各式商品
2.休息区用树干环绕起来，营造了大自然的氛围
3.螺旋楼梯状的展示架
4.木制格栅分隔了不同区域
5.胡桃木骨架船极具特色
6.整个空间纯净光亮

Besides these scenic and natural elements and finishes, glossy white surfaces characterise the space. The lacquered surfaces of the shelves, containers and platforms created interesting reflections, conferring a sensation of extreme purity and lightness to the space. The lighting project was conceived as an imitation of natural light. A careful layout of light cut on the ceiling, niches lit from within and skylights on the first floor make the space itself seem to radiate light.

The ground floor opens up with the entrance hall offering various services: fidelity, the pay desk and the bar, where the retailer can be welcomed, informed and escorted to the selection accessories. These service areas were separated from the exhibition space by walls made of walnut rods, creating a "transparency-opacity" effect to offer privacy without being a real barrier. The perspective of the walls at the entrance direct the attention to the centre of the space where an offset and scenic staircase leads to the upper level with the five showrooms, situated beyond another room-divider made with walnut rods.

The showroom of each brand is characterised by one key element: long legged pink chairs for Martissima, a castle of glossy red boxes for Krizia Poi, a big black mirror frame for roccobarocco, a never ending spiral staircase for Mario Valentino and, last but not least, a game of plates on various heights for a brand yet to be introduced.

每个品牌的展览室都各有特色：Martissima的粉红长腿座椅、Krizia Poi的红盒子城堡、roccobarocco的巨大黑色镜框、Mario Valentino的无限螺旋楼梯以及未知品牌的不同高度的飞盘。

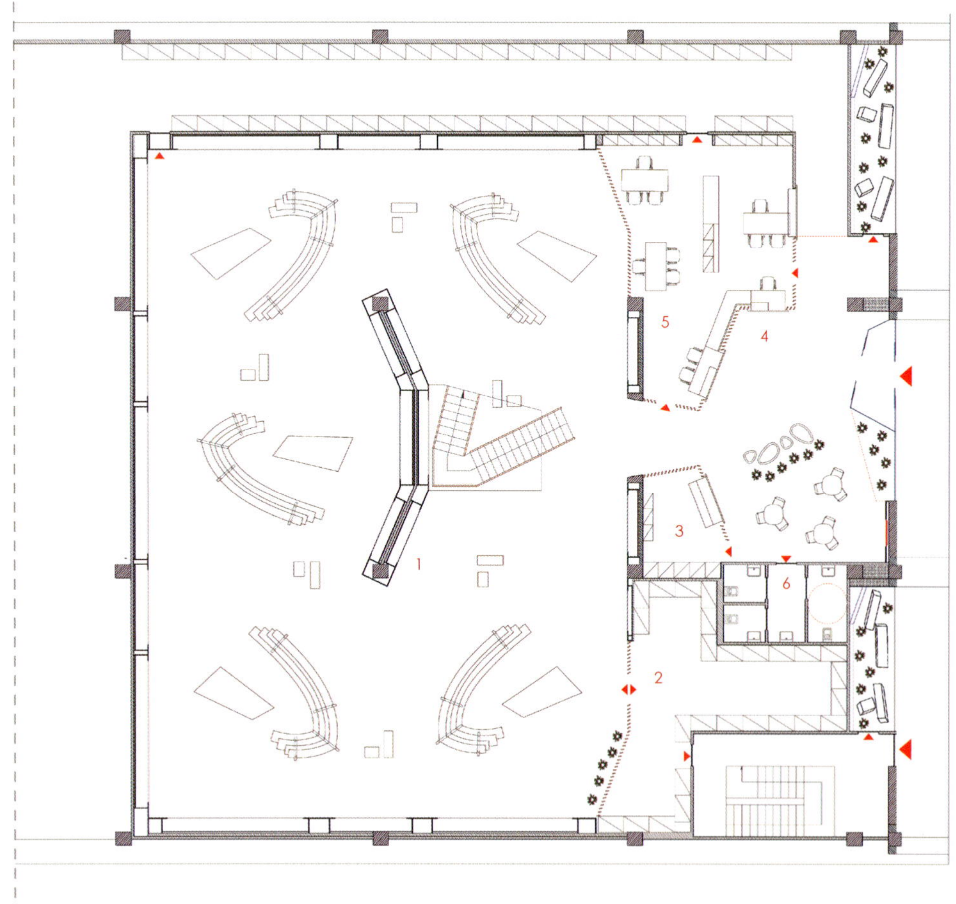

1. Exhibition area 1.展览区
2. Outlet 2.出口
3. Bar 3.吧台
4. Hall 4.大厅
5. Fidelity 5.洽谈室
6. Toilet 6.洗手间

3

4

116 - 117

5

6

1

Hermès Rive Gauche

爱马仕巴黎左岸店

Location:
Paris, France

Architect:
RDAI

Photographer:
Michel Denance

Completion date:
2010

项目地点：
法国 巴黎

设计师：
RDAI设计公司

摄影师：
米歇尔・德南斯

完成时间：
2010

At the foot of an elegant apartment building from the mid 1930s, the façade of the Hermès store is discreet. An entrance portico in the centre between two windows, nothing to hint at the surprise awaiting once through the doors.

The entrance is like a lightwell overturned, horizontal, which attracts one irrevocably towards the light at the back, towards what was the Lutétia swimming pool. The entrance to the store must function like a delicious trap into which the visitor lets himself slide, from crossing the threshold of the doors on the street until he reaches the swimming pool and its strange inhabitants, the huts. To guide him, the perspectives are accentuated and modified by an imperceptible contraction, rather like the sides of the Médicis fountain in the Luxembourg garden. The lightly inclined ceiling, the walls curved and leaning inwards, covered with oak laths leave recesses open as if floating in matter.

There are four pavilions with an organic design, in which some will recognise familiar forms from the plant or animal world, or from childhood... Others will liken these huts, which occupy the volume of the swimming pool, to the nests of tisserin birds. These pavilions of different form and dimensions were constructed in ash wood. They are self-supporting structures that rest on a system of woven wooden laths (profile

坐落在一座优雅的住宅楼楼下的爱马仕专卖店外墙十分低调。两扇窗户之间是一个拱门，令人难以想象其内部令人惊喜的装饰设计。

店铺入口仿佛一个倒转的水平采光井，吸引着人们走向后部的游泳池。入口仿佛一个诱人的陷阱，吸引着客人深入探索——从街道进入店铺，走向泳池，到达奇怪的木屋。远景有一种难以名状的吸引力，就像卢森堡花园里的梅迪西斯喷泉一样。微微倾斜的天花板、弯曲的墙壁和前倾的橡木条都吸引着人们向前。

四个展馆的有机造型设计来源于动植物世界或童年时代的想象。其他的则像木屋一样占据着泳池的四周，像鸟巢一样。这些造型各异的展馆由白蜡木制成，它们凭借编织的木条系统达成了自我支撑。每个木屋的档案和复杂的3D绘图都由电脑脚本制作。

展馆高9米以上，依次倾斜，仿佛被天窗所吸引一样，里面展示着爱马仕的产品。它们看起来似乎采用了地面照明，更具游动。第四个木屋像倒下了一样，与通往泳池的楼梯并列，形成了入口和泳池开放空间的连接线。

2

1. The store has an open and clean layout
2. The four main pavilions consist of curved oak laths
3. The chairs and tables are also in wood
4. A corner of the store
5. The arc-shaped counter
6. The display area is mysterious and inviting
7. The upward pavilions seem to be attracted by the skylight
8. The streamlined staircase

1.店铺内布局开阔整洁
2.弯曲倾斜的橡木条构成了四个主要展馆
3.中间摆放的桌椅也采用浅木色系
4.店内一角
5.弧形的橱窗柜台
6.鸟巢状的展示区充满了神秘感和吸引力
7.向上延伸的展馆仿佛被天窗所吸引一样
8.流线型的楼梯设计

6 centimetres x 4 centimetres) with a double radius of curves. The documentation and three-dimensional drawing of the complex geometry of each hut was made possible by the computer script written for each one of them.

Rising to more than 9 metres in height, they lean progressively, as if attracted by the skylights. The huts house the Hermès collections. They seem to have simply alighted on the ground, lending the project its nomadic dimension. The fourth hut, which appears to be lying down, lines the staircase that naturally leads the visitor towards the pool and forms the link between the entrance and the open space of the swimming pool.

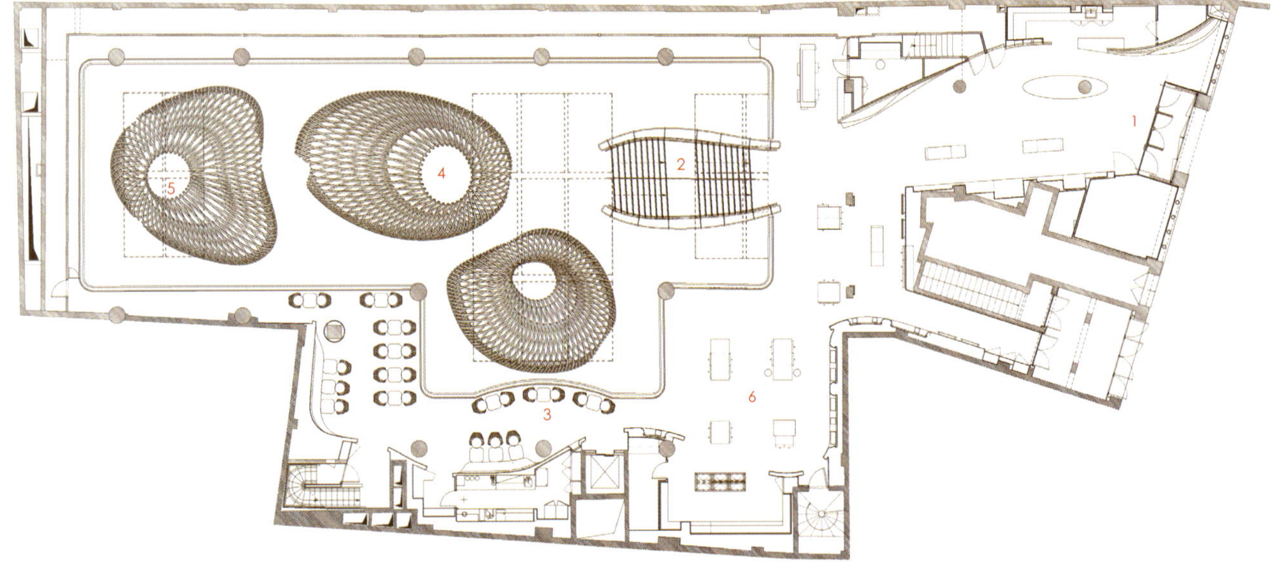

1. Entrance
2. Stair
3. Café area
4. Dining area
5. Clothes display area
6. Library

1.入口
2.楼梯
3.咖啡区
4.就餐区
5.服装展示区
6.图书室

3

4

5

6

7

8

1

Kaloo (Prince Building)

卡路（太子大厦）

Location:
Hong Kong, China

Designer:
Axiom Design Partnership Ltd.

Photographer:
Roger Ho

Completion date:
2007

项目地点：
中国 香港

设计师：
艾森设计顾问事务所

摄影师：
何剑强

完成时间：
2007

Kaloo is an international brand, which was created in 1998 by a French national Mr. Eric Panciulo. This brand was distributed into 35 cities worldwide, such as Paris, Munich, Madrid, Istanbul, Tokyo, Hong Kong and Taiwan.

In respecting to the original spirit of the Kaloo, and adopting the European culture, the concept was inspired by "Love, Care, and Happiness".

Because the design would be implemented worldwide, the given objective was to create a retail image with flexible system, to be easy adopted and easy appreciated by customer. Therefore, a warm, homely, harmonious and welcoming atmosphere has been reflected by a modern way of mixing use of French Oak and white wall. The design also integrated with two parties, modular system for easy and flexible showing Kaloo product, such as soft toys, furniture and textile items. And soft elements have been created for identity enhancement.

卡路是一个国际品牌，由法国人埃里克·潘西罗先生创立于1998年。该品牌遍布全球的35个城市，包括巴黎、慕尼黑、马德里、伊斯坦布尔、东京、香港、台湾等。

设计理念是“关爱与幸福”，尊重卡路的品牌精神的同时，融入了欧洲文化。

由于设计将在全球范围内实施，设计师的目标是建立一个灵活的零售店形象，容易被不同的客户接受。设计师采用法国橡木和白色的墙壁，营造了温馨、亲切、和谐、热情的氛围。店内还安排了两间展示室，方便和灵活地显示卡路的产品，包括毛绒玩具、家具、纺织品等。同时以布艺装饰强调品牌形象。

2

1. Passers-by can see the warm interior design through the display windows
2. The wood floor and walls create a warm and natural atmosphere
3. The goods in display are quite cute
4. The lighting fixtures set in the walls cast a soft and warm light
5. The brand's concept "Love, Care, and Happiness" is shown in the store design

1.透过玻璃橱窗可以看到店内温馨的设计风格
2.木制地板和墙壁营造了亲切自然的氛围
3.商品展示十分可爱
4.镶嵌在墙壁里的灯光柔和温暖
5."关爱与幸福"的设计理念在店内得到完美体现

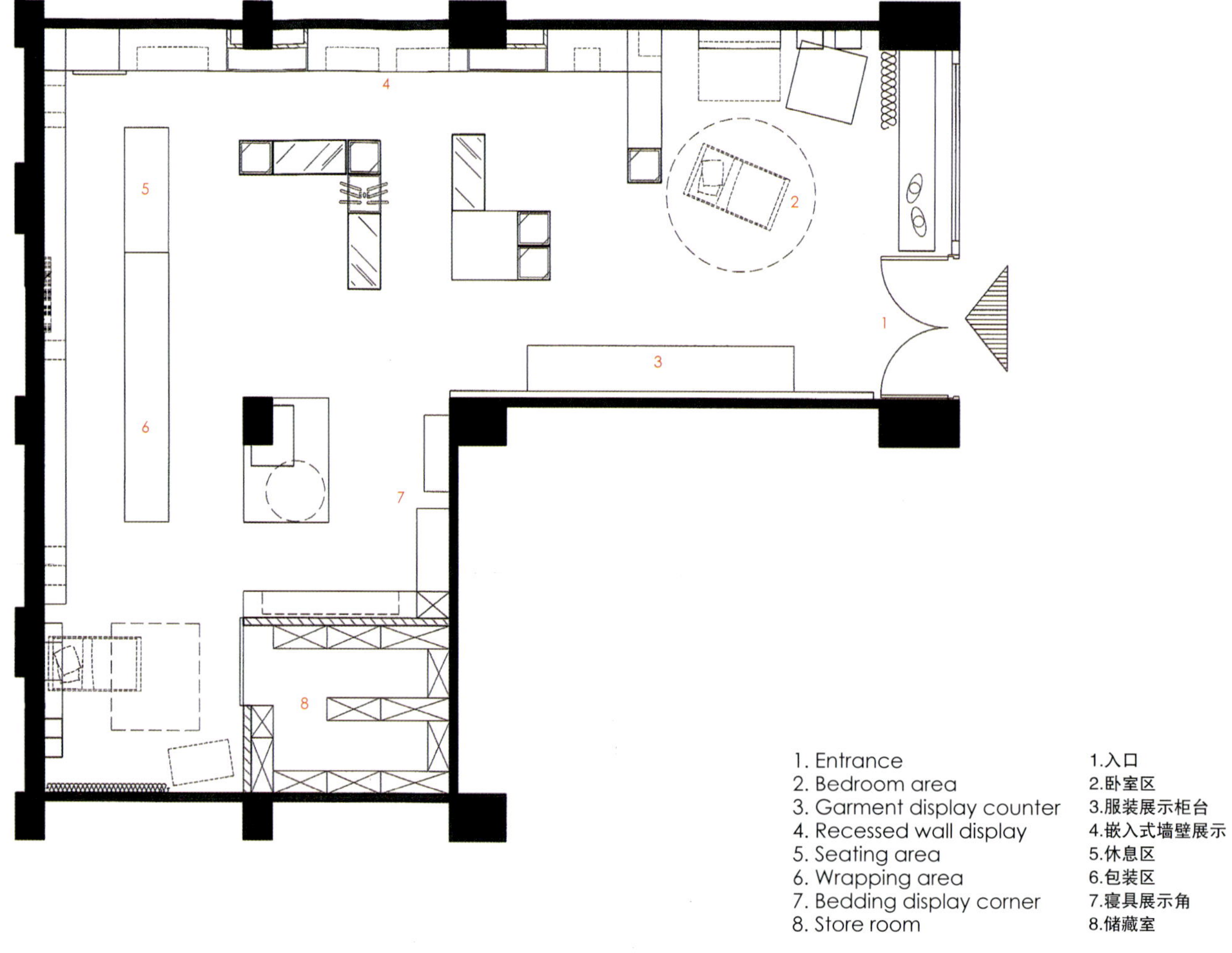

1. Entrance
2. Bedroom area
3. Garment display counter
4. Recessed wall display
5. Seating area
6. Wrapping area
7. Bedding display corner
8. Store room

1.入口
2.卧室区
3.服装展示柜台
4.嵌入式墙壁展示
5.休息区
6.包装区
7.寝具展示角
8.储藏室

Kaloo

Kaloo
Kaloo

1

Komplementair Accessories

互补精品配饰店

Location:
Zurich, Switzerland

Designer:
Aekae

Photographer:
Nico Schaerer

Completion date:
2010

项目地点：
瑞士 苏黎世

设计师：
Aekae设计公司

摄影师：
尼克·沙勒尔

完成时间：
2010

Aekae has designed the interior for Komplementair, a boutique devoted to the world of accessories. The 140-square-metre location required a very unique approach; in the old railroad viaduct, running through Zurich's rapidly changing Kreis 5, the original stonewalls are part of the interior. "Komplementair" refers to the accessories complementing an outfit; this theme was used on a bigger scale for the interior design as well.

The goods are not as usual displayed in categories, but in assorted groups of different items. To accomodate the wide range of objects, from small jewellery to weekend bags, various secondhand furniture pieces were compiled to bigger structures, offering various plattforms and niches of different sizes. The heart of the installation on the ground floor is a Steinway Grand Piano, which is still played on at special occasions.

All surfaces were covered with custom fitted lacquered MDF to create a neutral backdrop for the products on display. After the search for furniture pieces this concept required a lot of on-site-work, experimenting with different layouts. On the upper floor the same approach was applied to the floor, using reclaimed hardwood parquet to create a patchwork.

The lighting system was specifically developed, to create additional display possibilities above

Aekae设计公司为互补精品配饰店进行了室内设计。140平方米的空间地处苏黎世克莱斯5区的铁路高架桥下，原有的石墙是室内的一部分。“互补”意味着配饰让整体更加完整，这一主题被融入到室内设计之中。

商品并没有像寻常一样分门别类摆放，而是被分成一堆堆混杂的种类。为了展示五花八门的商品（从小珠宝到旅行袋），大量的二手家具被摆放在室内，提供了各种各样的平台和尺寸不一的壁龛。一楼的正中央是一架斯坦威大钢琴，可以在特殊场合演奏。

所有的桌面上都覆盖着定制的喷漆纤维板，为产品提供了一个干净简洁的背景。除了家具的选择之外，项目还包含很多现场的工作，进行不同的布局。二楼的地面上采取了再生硬木拼花地板形成了拼嵌图案。

灯光系统经过了特别开发，在家具之上增添了新的展示平台。灯具上的黄铜杆可以悬挂物品，灯罩可以通过绳索简单地悬挂在杆上，任意调节高度。收银台和洗手间都采用了简洁的黑色模块，与白色的地板和石头墙壁形成了对比。

店主要求室内设计采用既激烈又折中的风格，即真实氛围中的精品店：粗糙的墙壁与精细的木工工艺和精美的细部处理相得益彰。店铺的标识（由Aekae和费边·利恩伯格共同设计）也传达了这种美感。

1. The entrance of the store
2. The suspended lamps are simple but unique
3. The display area with various products
4. The vintage furniture adds retro atmosphere for the store
5. Black curtains emphasise the brightness of display area
6. The primitive stone wall makes a contrast with the fine products

1.精品店入口
2.交叉悬挂的吊灯简单而又独具特色
3.种类丰富的商品展示区
4.半旧的家具增添了店内的怀旧感
5.黑色的幕帘更凸显了明亮的展示区
6.古朴的石墙与精美的商品形成鲜明对比

the furniture. It combines the brass rods for hanging items, with height adjustability of the lampshades by simply wrapping them around the rods. The infrastructural elements, counter and bathroom, are simple black blocks, standing on the white floor, which contrasts to the stonewalls of the viaduct.

The owner asked for a strong and eclectic characteristic; the result is a boutique with an authentic vibe, where the raw walls are complemented by the fine carpentry work, and refined details. Also the identity, created by Aekae's studio associate Fabian Leuenberger, conveys this sense of aesthetic.

1. Entrance 1.入口
2. Display stand 2.展示柜台
3. Toilet 3.洗手间

3

4

5

6

1

L'Eclaireur Paris

侦察兵服饰巴黎店

Location:
Paris, France

Designer:
Studio Arne Quinze

Photographer:
Eric Laignel

Completion date:
2009

项目地点：
法国 巴黎

设计师：
阿纳·奎兹工作室

摄影师：
埃里克·莱格尼尔

完成时间：
2009

The shop is as much an art installation as a retail outlet. Dominating the sales floor is a Quinze-style freestanding slatted form that looks a bit like a spindly three-legged dinosaur. Surrounding walls were covered with recycled wooden slats, two tons of them, nailed together with small panels, aluminium printing plates, cardboard scraps, and other bits. "Essentially, it's an amalgam of trash," but artfully arranged, composed on the spot, piece by piece, in exactly the same way Quinze's sculptures grow gradually and organically.

Surfaces were then thickly coated with automotive paint in a silvery gold. The paint was allowed to form drips and bubbles, making the already irregular surface even more highly textured, while changing ambient light adjusts the effect. Lighting is a combination of natural and artificial, as modular LEDs supplement the sales floor's skylit roof. Mirrors, scattered around, reflect the iridescent walls. At times, the entire space appears to be in motion.

Walls are punctuated by square and rectangular niches. Smaller ones, even those in the fitting rooms, are used to display accessories as well as exotic artifacts. (Niches holding jewellery have lockable glass doors.) The niches animate the walls, but they're not alone in doing that. Interspersed with them, small video screens can be programmed to

比起零售店，这家店铺更像一件艺术装置。销售区域最显眼的便是奎兹风格的板条造型雕塑，看起来像一只三条腿的恐龙。墙壁上附带着回收木板条（总重量达2吨），它们与印刷铝板、硬纸板和其他东西钉在一起。这些东西本来是一堆垃圾，但是经过艺术处理，变成了具有雕塑感的艺术品。

墙面上刷着厚厚的银色汽车涂料。在涂抹过程中，涂料形成了水滴和气泡，让本来就杂乱的表面更加具有质感，同时也改变了灯光效果。灯光设计结合了自然采光和人造照明，模块化LED灯光填补了销售区天窗的自然采光。散布在空间里的镜子反射着闪光的墙壁，让整体空间充满了动感。

墙壁里嵌着正方形和矩形的壁龛，里面展示着配饰和具有异国情调的工艺品（展示珠宝的壁龛里有玻璃罩）。壁龛和滚动播放flash的小型屏幕让墙壁鲜活了起来，营造了活跃的氛围。

墙壁的下部是稍大的壁龛，里面悬挂着服装。这些壁龛隐藏在电子控制面板上，可以向外或向上开合。售货人员为准买家开启衣橱，营造出惊喜感。每个衣橱代表着一位设计师、一种风格或是一个种类。从侧墙几乎看不到的高大的双开门打开后会变成传统式女装三开镜。而另一个隐藏门里面是更大的惊喜——一个名为Echo & Narcisse的视听装置。

2

1. Walls are punctuated by square and rectangular niches
2. Walls with automotive paint have a unique texture
3. The columns with waste materials become art works
4. The deck chair echoes with the mysterious interior design
5. Modular LEDs supplement the sales floor's skylit roof
6. A slatted art work looking like a dinosaur

1.店内墙壁上镶嵌着方形壁龛
2.刷有汽车涂料的墙壁充满质感
3.废旧材料附着的柱子变成了艺术品
4.个性躺椅十分切合室内的神秘风格
5.模块化LED灯光填补了销售区天窗的自然采光
6.恐龙造型的木板条艺术装饰

flash branding images – to support a perfume launch, for example, or just to create a mood.

Along the bottom of the walls, much larger niches hold hanging clothes. These niches are hidden behind electronically controlled panels that swing out and up, pulling the racks forward in the process. Salespeople open the closets for prospective buyers, making the wares a special surprise. Each closet is devoted to a single designer, style, or type of merchandise: Dries Van Noten, Comme des Garçons, leather-wear. Tall double doors, almost invisible on a sidewall, open to become a traditional couturier's triptych mirror. And another set of hidden doors opens to reveal perhaps the biggest surprise, an audiovisual installation titled Echo & Narcisse.

1. Entrance hallway 1.入口走廊
2. Central room 2.中心厅
3. Secret room 3.密室
4. Dressing room 1 4.更衣室1
5. Dressing room 2 5.更衣室2
6. Corridor 6.走道
7. Multi room 7.多功能室
8. Storage 8.储藏室
9. Event room 9.活动室

3

4

5

6

1

Louis Vuitton New York Office and Showroom

路易威登纽约办公室和展示厅

Location:
New York City, USA

Designer:
ikon. 5 architects

Photographer:
James D'Addio

Completion date:
2009

项目地点：
美国 纽约

设计师：
ikon5建筑事务所

摄影师：
詹姆斯·达迪奥

完成时间：
2009

The project for Louis Vuitton is an interior renovation of the eighth and ninth floor of the flagship store at 1 East 57th Street, New York. The design incorporates a neutral and minimal palette inspired by the colours, textures and patterns of the Vuitton design tradition. The bright white surfaces of the walls and ceiling and interior glass partitions assist in harvesting light from the perimeter offices and delivering it to the centre workstations. Luxurious wood floors were used in the showroom to convey a warm rich experience that is both modern and timeless.

The project involved the complete interior renovation of a pre-existing office space. The space was a labyrinth of walls and low ceiling. To mitigate this closed and tight feel, the designers arranged the offices and conference rooms around a central open workstation environment. Interior partitions between the open workstations and the offices and conference rooms are glazed to permit natural daylight and views to penetrate deep into the central portion of the floor plate. An internal stair permits a physical connection between the eighth and ninth floor which house the Marketing and Public Relations offices, respectively, so that these two interdependent departments can easily have access to each other.

路易威登办公室位于纽约东1区第57号街的旗舰店，其室内设计受路易威登传统色彩、材质和图案的启发，体现了中性、简单的色调。亮白的墙壁、天花板和透明隔断有助于将周边的光线传递到中央的工作台。展示厅里奢华的木地板传递了一种温暖的感觉，现代又经典。

项目包括办公空间的整体翻新。原有的空间杂乱、天花板过低。为了缓解这种封闭的紧张感，设计师将办公室和会议室围绕中央开阔的工作台展开。办公台和办公室、会议室之间的玻璃隔断能有效地让自然采光渗透到室内中央。内部楼梯将8、9楼的市场部和公关部门联系在一起，让这两个部门可以自由交流。

室内细节和装饰都极其简单，强调了路易威登的品牌美学。例如：玻璃隔断采用了对接工艺，将支撑玻璃墙的垂直结构减到最小，让墙面尽量透明。展览厅位于第5街和第57街转角的8楼，里面展示着最新的路易威登系列商品。

整体翻新项目包括市场部和公关部办公室、会议室、产品展示厅、更衣室和仓库。

2

1. The open store positioning
2. The interior lighting is bright and soft
3. Glass walls divide the space into different areas
4. The display of products
5. The staircase connecting two areas
6. Upper-floor working area
7. The store has a good daylighting

1.宽敞的店内布局
2.室内光线明亮柔和
3.玻璃墙壁分隔各个空间
4.商品展示
5.连接两个区域的楼梯
6.楼上的工作区
7.店内采光良好

The interior details and finishes are designed to minimally express the architectonic connection between surfaces and to highlight the Louis Vuitton aesthetic. For example, the interior glass partitions are butt glazed to minimise the vertical structure needed to support the glass wall and allow the wall to be as transparent as possible. The showroom is located at the corner of the 5th and 57th street on the eighth floor. This retail space displays the most current pieces in the Vuitton collection.

The interior renovation of the Louis Vuitton New York offices includes the offices of the Marketing Department, office of Public Relations, shared conference rooms, product showroom, changing rooms and product storage.

1. Showroom 1.陈列室
2. Conference room 2.会议室
3. Kitchen 3.厨房
4. Packing 4.包装室
5. Media 5.媒体室
6. Office 6.办公室
7. Lobby 7.大堂
8. Changing 8.更衣室

3

4

5

6

7

1

V2K Designers, Nisantasi

尼桑塔希V2K旗舰店

Location:
Istanbul, Turkey

Designer:
Seyhan Ozdemir & Sefer Caglar

Photographer:
Ali Bekman

Completion date:
2009

项目地点：
土耳其 伊斯坦布尔

设计师：
赛伊汉·奥兹德米尔&瑟夫尔·卡格拉尔

摄影师：
阿里·贝克曼

完成时间：
2009

V2K Designers – a subsidiary brand of Turkey's premiere luxury fashion house Vakko, is a multi-label apparel store, displaying collections from the likes of Alexander Wang, Erdem, Hussein Chalayan, Rick Owens, Band of Outsiders and etc. Founded in the year 2000, the brand is known for carefully mixing up-and-coming designers with some established ones that have an edgy look. The store also carries non-fashion items, such as books on design, designer headphones and cool city bicycles.

Opened in late 2009, in Nisantasi – one of Istanbul's main shopping hubs, the new V2K Designers Flagship Store was Designed to reflect the fashion-forward mood of the brand, and bring together fashion and innovation. The store welcomes you with an angular entrance, which radically contours the borders of the store and separates it from the timeworn façade of the building rising above. The same angular entrance opens into a lit path that runs up to the mezzanine floor, enabling a large window space for more theatrical displays. On the left is a catwalk-like area where mannequins dressed in the latest fashion stand in line to attract the attention of the passer-by. Above them is a grid-like wall of light bulbs that runs down the side of the store, providing a banner space that can easily be customised by changing the slot of the bulbs. This 258-square-metre V2K Designers store has two floors. The main floor entirely reserved for women's wear

V2K是土耳其奢侈服饰店Vakko的附属店，经营各大品牌服装，其中包括：亚历山大·王、艾尔丹姆、侯赛因·卡拉、瑞克·欧文斯、旁观者等品牌。V2K品牌创立于2000年，以完美的结合新兴设计师和知名设计师的作品为闻名。V2K也经营非时装物品，如设计图书、耳机和自行车等。

V2K旗舰店于2009年开张，位于伊斯坦布尔的主要商业区尼桑塔希，反映了该品牌的时尚感，将时尚和创新融为一体。棱角分明的入口将店铺与上方古旧的建筑分割开来。入口正对通往阁楼的光道，为橱窗展示提供了大片空间。左侧是一块类似T台的区域，一排身着最新时装的模特吸引着行人的目光。模特上方是一面装满灯泡的网点状墙壁，可以用灯泡摆成任意的标语。258平方米的店铺空间分为两层，主楼层经营女装和配饰，阁楼经营男装和女式晚礼服。

店铺的室内设计显示了设计师的风格，极具创意。整个空间几乎是全白的，其目的是让人们将注意力集中在展示的商品上。除了天花板上的灯光外，其他的灯光系统都是隐蔽式的，突出了服饰，而中心的木板圆柱打破了室内的单一感，增添了温馨的氛围。

V2K旗舰店不仅是一家零售店，而且是一个创意空间，经常进行设计、艺术和摄影作品展览。因此，定制的不锈钢货架栏杆可以进行改装，仅需要小小的调整，就能让室内空间截然不同。

2

1. White backdrop can highlight the products in the store
2. Besides products, the store also blends in cultural elements
3. The interior structure is quite innovative
4. The design is full of uniqueness and fashion
5. The lighting hidden in the wall is pretty soft
6. The display of the products

1.白色背景更好的突出了店内的商品
2.除了商品店内还融入了文化元素
3.店内构造充满创新
4.极具个性与时尚感的设计
5.隐蔽在墙壁里的灯光十分柔和
6.商品展示

and accessories, while the mezzanine floor houses men's collection opposite to women's eveningwear.

Incorporating the styles of the designers, the store's interior is innovative and almost all white to put the focus on the collections on display, rather than surpassing them. In addition to the ceiling lights, the lighting system hidden behind departments also put the limelight on clothing, while central columns covered in wood panels break the monochrome look of the interior to add warmth.

More than just a retail shop, the new V2K Designers Flagship Store also doubles as a creative space, hosting design, art and photography exhibitions on a regular basis. Therefore, the custom-made stainless steel bars holding racks are also designed to allow customisation. The overall result creates an impression of a space that has some big changes to the interior with actually only a small amount of alteration.

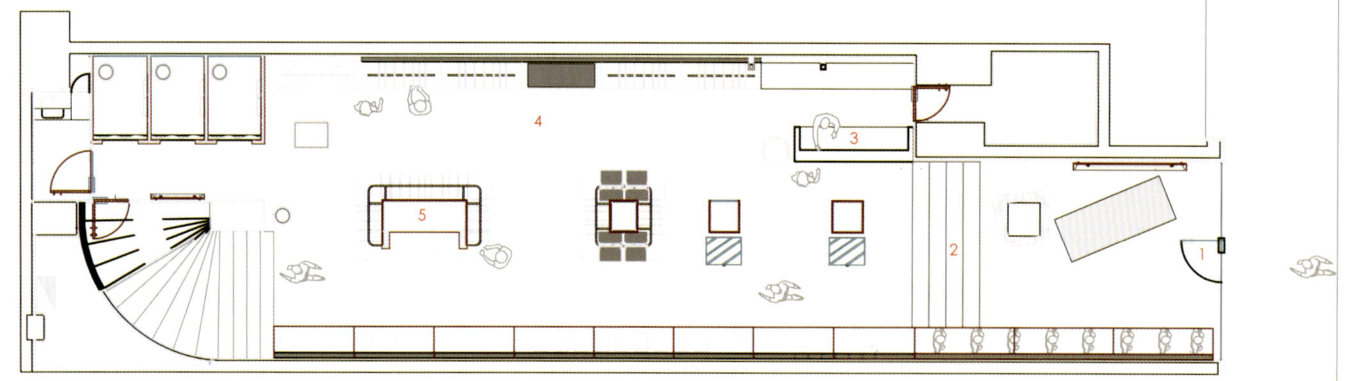

1. Entrance 1.入口
2. Stairs 2.楼梯
3. Counter 3.柜台
4. Main sales room 4.主销售区
5. Sofa 5.沙发

3

MAISON
MICHEL
PARIS
SINCE 1936, THE MAISON MICHEL HAS
CONTINUED TO PERFECT ITS ART CREATING
HATS FOR THE MOST PRESTIGIOUS STUDIOS.
IT IS RECOGNISED TODAY FOR ITS VAST
CAPACITY TO EXPLORE SHAPES AND
MATERIALS.
LAETITIA CRAHAY, HEAD OF ACCESSORIES
AND JEWELLERY AT CHANEL AND ARTISTIC
DIRECTOR OF THE MAISON MICHEL, HAS
DESIGNED SINCE 2006 READY-TO-WEAR HATS
AND HAIR ACCESSORIES. FRESH AND
SEDUCTIVE, WITH A STRONG IDENTITY, AND
PERFECTLY IN THF SPIRIT OF THE TIMES,
HER CREATIONS HAVE ALREADY BECOME
ESSENTIALS.
BETWEEN AN ULTRA-FEMININE REFINEMENT
AND A ROCK SPIRIT, THE FALL-WINTER
2009/2010 COLLECTION REVISITS A LARGE
REPERTOIRE OF CLASSIC SHAPES, FROM
THE MELON HAT TO THE BOATER AND THE
TRILBY. ACCESSORIES, HEADBAND
JEWELS FOR THE HAIR UNITE FE
LACE, PEARLS OR CHAINS AND B
LINES WITH HUMOUR.
IN THIS COLLECTION OF PORTRAI
FRIENDS OF LAETITIA CRAHAY POS
FRONT OF THE CAMERA OF KAR
LAGERFELD, AND MAKE THE PIECES O
THE FALL/WINTER 2009/10 COLLECTIO
THEIR OWN, ACCORDING TO THEIR
DIFFERENT PERSPECTIVES OF THE
WORLD AND TO THEIR ECLECTIC
PERSONALITIES.
MICHEL
CHAPEAU!
MAISON
PARIS

4

5

6

1

Van de Velde Showroom

凡德维尔德展示厅

Location:
New York City, USA

Designer:
LABscape Architecture

Photographer:
LABscape Architecture

Completion date:
2010

项目地点：
美国 纽约

设计师：
LABscape建筑事务所

摄影师：
LABscape建筑事务所

完成时间：
2010

The brand new showroom for Belgian lingerie firm "Van de Velde" opened its doors in March 2010 at the high pedestrian traffic intersection between the Madison Avenue and 33rd Street, on the ground floor of a historic building, in New York. The typical "Manhattan" open space needed to be divided into four different areas. The entrance, a closed office for one person – that could also be used as a meeting room – two work islands and a showroom area.

The creative idea was to use, as leitmotiv, lace design: its weave, its structure and the geometry arising from stretching and deforming it, moving away from the romantic, yet archaic and tried image, of lace in favour of a contemporary reinterpretation.

Using an inversion game, the structure becomes void and the void becomes matter. The result is a perfect balance between altered volumes, the linearity of the furniture and the interior envelope. The entrance is a passage area between the different zones. To the right, an almost evanescent glass cube creating the opportunity either for a workspace or a private meeting. The showroom area extends to the left with a wall made of 21 deformed cells that are used either to display or store merchandise. A volumetric progression of functional boxes. The contrasting colours, white and anthracite, offer the staging for the products.

2010年3月，比利时内衣公司"凡德维尔德"的全新展示厅在纽约繁华的麦迪逊大街和第33街交叉口处一座历史建筑的一楼开张。这个典型的"曼哈顿"开放式空间被划分成4个区域：入口、单人封闭办公室（也作会议室用）、两个工作台和一个展示区。

空间的设计极具创意地使用了蕾丝：它的波纹、结构和几何图案被拉长和变形，脱离了原有浪漫、古老的形象，被重新赋予了现代的诠释。

设计师采用了反转理念，让结构变空，让空变实。最终达成了各个空间、家具的线性结构和室内围度的完美平衡。入口是不同区域之间的走廊，向右是一个透明格子间，既可作为办公室，又可作为小型会议室。展示厅向左侧延伸，墙面上有21个变形格子，用来展示或储藏商品。这些格子排列成形，各具功能性。煤灰色和白色两种对比色为商品展示营造了背景舞台。

这个由具有意大利背景的纽约工作室LABscape所设计的全新空间打破了内衣店的传统设计，具有全新的现代风貌。

2

1. The grid display windows with geometric patterns
2. The design is novel and fashionable
3. Different areas transit naturally
4. The transparent office
5. A general view of the store
6. The white service stand
7. The display of the products

1.几何图形构造的格子橱窗
2.设计新颖时尚
3.各个区域过渡自然
4.透明格子间
5.店内空间纵览
6.白色工作台
7.商品展示

It is a new space breaking away from the design conventions of lingerie's retail environments with novel and contemporary look developed by Italian-New York based studio LABscape.

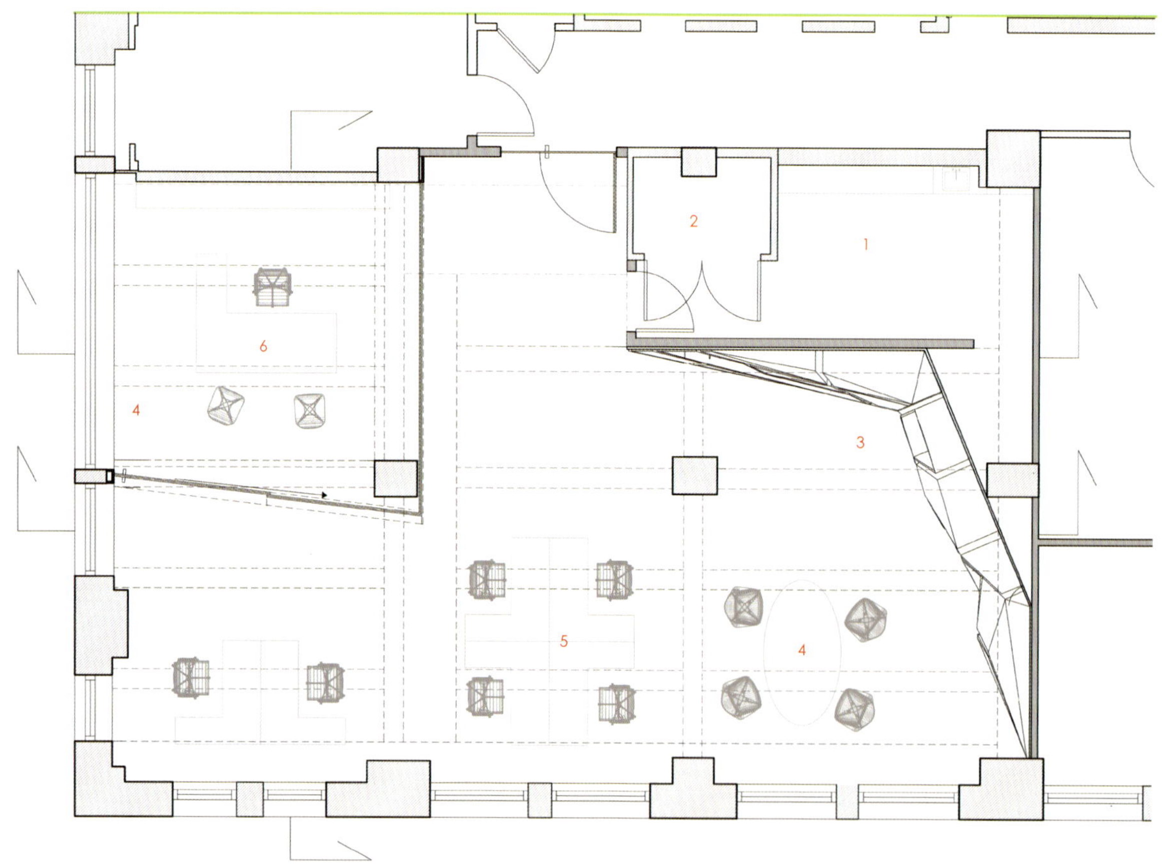

1. Fitting room 1.试衣间
2. Air-conditioner room 2.空调房
3. Product presentation 3.产品展示区
4. Meeting room 4.会客室
5. Work table 5.工作台
6. Enclosed workspace 6.封闭工作区

3

4

5

6

7

1

BRIO Store

布里奥亲子店

Location:
Stockholm, Sweden

Designer:
Urban Design

Photographer:
Per Kristiansen

Completion date:
2008

项目地点：
瑞典 斯德哥尔摩

设计师：
城市设计事务所

摄影师：
波尔·克里斯汀森

完成时间：
2008

In 2006 Urban Design completed a new headquarters for BRIO in Malmö. Through this project the designers discovered the playful and sophisticated world of the company. BRIO's first own store is situated in the central parts of Stockholm, adjacent to the fashion street, a position well suited to the brand and its vision – to inspire and support the modern parenthood.

The customer is invited as a guest in the store, which is also a service spot for BRIO's products. If your trolley is broken, you are welcome to borrow another, while your own one is repaired. You also have the opportunity to feed your baby or use the bathroom. Outside the store there is a parking for trolleys, with a possibility to lock it up. The main sign is a 3D logo, borrowed from the headquarters in Malmö, in one of the large shop windows. A second sign, the red glass box, is visible from all directions from its position on the corner.

The store is located in a glass corner of a building from the modernistic era. The designers wanted to keep as much of its original atmosphere as possible, making BRIO a guest on this location. Under the carpets of the latest tenants, the designers found a nice worn wooden floor which they uncovered. One part of the floor was painted white, to establish a calm spot for the displayed toys. The wall behind was painted red, with a shelving system,

2006年，城市设计事务所为布里奥公司在马尔默设计了新总部。通过这个项目，设计师发现了该公司活泼而不失成熟的氛围。布里奥公司的第一家亲子店坐落在斯德哥尔摩市中心，紧邻时尚街区。这一位置十分符合布里奥的品牌价值和理念——鼓励并支持现代亲子关系。

进入店铺的消费者会有宾至如归的感觉，这是BRIO产品的服务亮点之一。当你的婴儿车坏掉并进行修理时，你可以在店内借用一个。你还可以在洗手间给孩子喂奶。店外有一个专门的婴儿车停车场，还配备了车锁。橱窗上的3D标志借鉴了布里奥公司总部的设计。另一个符号是转角处的红色玻璃盒子，这一设计让人从四面八方都可以看到它。

店铺设在建筑的一角，外部全部采用了玻璃造型。设计师试图保留建筑的原有氛围，让布里奥亲子店看起来像客人一样。设计师移除了地毯，露出了做旧的木地板。店内一部分地面被漆成了白色，上面放置了家具店常用的货架结构。这个区域摆放着低矮的桌椅，它们和吊灯一起都是阿尔瓦·阿尔托设计的。孩子们可以在这里玩耍。对面的展台上摆放着布里奥最新的婴儿车。正中狭窄的走道尽头是一面镜子，你可以试试婴儿车的“上身效果”，就像试穿衣服一样。店铺内部是一块抬高的区域，展示布里奥的系列家居用品。这三个区域共同构成了店铺的主要功能区——游戏区、驻留区、走动区。

Jag
rymmer!
BRIO

1. The windows are clear and bright
2. The play area is lovely and lively
3. The display of a trolley
4. A general view of the store
5. A corner of the store
6. The bathroom has a beautiful and warm atmosphere

1.玻璃窗明亮通透
2.游戏区设计得活泼可爱
3.婴儿车展示
4.店内总览
5.商品展示一角
6.卫生间也设计得十分美观温馨

borrowed from home furnishing rather than store systems. This area includes low tables and chairs, together with light pendants, all classics by Alvar Aalto. This is an interior playground for visiting kids. On the opposite side, a series of podiums exposes the range of BRIO's latest trolleys. The central path constitutes a catwalk which ends up in a mirror that gives you an opportunity to "try your trolley on", just like a new suite or a dress. The inner department is an elevated space for home interiors, exposing BRIO's range of furniture. These three areas constitute the main product sections – to play, to stay and to go.

1. Stroller parking
2. To borrow - prams/strollers you can borrow if yours is being fixed
3. Now - New objects from the BRIO collection
4. To go - Display of prams/strollers
5. To play (toys)
6. Nurse - Chair for nursing babies
7. Sit - Display of high-chairs for children
8. Look - Mirror on the wall on the "catwalk" for prams
9. Coffee/juice
10. To be - Display of furniture
11. Toilet

1.婴儿车停车处
2.借车处——在婴儿车修理时，顾客可以租借婴儿车
3.最新产品——布里奥新品
4.婴儿车展示区
5.玩具区
6.婴儿看护区
7.座椅区——展示儿童高脚椅
8.婴儿车使用体验区
9.咖啡/果汁区
10.家具展示区
11.洗手间

BRIO Go
BRIO Go

4

5

1

Casa Natura

自然之家

Location:
Sao Paulo, Brazil

Designer:
Epigram + FGMF

Photographer:
Fran Parente

Completion date:
2009

项目地点：
巴西，圣保罗

设计师：
妙语设计 + FGMF设计事务所

摄影师：
弗兰·帕兰特

完成时间：
2009

When architecture, communication and branding come together with a common goal, spaces turn into place: place of sensory experiences, meeting place, a place to be, finally, a place of expanding awareness and knowledge. When Natura hired Epigram group and FGMF to develop the so-called "Casas Natura" (Natural houses), places dedicated to brand experiences, experiences with products and training its sales force, or even to final consumers, all this was taken into account.

In this sense, the architectural development has become a truly integrated design, it was made a consulting brand positioning, conveying the values and principles of Natura into space. Thus increased by studies on the flow and functional zoning, from a discipline known as visual merchandising, to achieve the architectural design. All this in an interdisciplinary process, in which there is no distinction between architecture and complementary projects, showing a systemic and open architectural design.

On the first phase of implementation, five houses were designed and built on the Greater Sao Paulo area, at different neighbourhoods. Appropriating existing buildings and entirely transforming them, these new spaces are intended to receive Natura's consultants for training, introducing them to new releases of the brand, and may also be a place for them

当建筑、沟通和品牌以共同目标融合在一起时，空间就变成了地点——感官体验的地点、集会的地点、停留的地点、扩充知识的地点等。当自然公司雇用妙语集团和FGMF设计事务所来开发“自然之家”时，地点意味着品牌体验。“自然之家”可供顾客体验产品，供公司培训销售人员，还可提供销售空间。

从这种意义上讲，建筑的开发成为了一项真正的综合设计。它显示了品牌定位，传达了品牌价值和理念。设计师通过视觉营销法则，研究了客流量和功能区，实现了建筑设计。这是一个跨学科的设计过程，建筑和配套工程之间没有了界限，形成了系统而开放的建筑设计。

在第一阶段，设计师在大圣保罗区域内设计建造了五座建筑。自然公司的咨询顾问可以在这些地点接受培训，了解新产品，并带领他们的客户前来体验新产品。这五个地点拥有相同的功能区：产品展示和体验区、2-3间礼堂、咖啡馆、行政区、仓库和辅助区。

建筑和家具的设计相互合作，形成了兼具美感和功能性的凝聚空间。每间“自然之家”都具有相同的元素，如入口隧道、木质百叶窗、水镜和室内外花园，便于客户辨识。这五个项目是“自然之家”进行全国性扩张的初期工程。

2

1. It is a spacious space
2. The wood floor and ceiling echo with the store's natural theme
3. The delicate and peaceful hallway
4. Aerial view of the hall
5. All areas in the store are combined with natural elements
6. The open space provides customers with better shopping experience
7. Experiencing and relaxing area

1.店铺空间十分宽阔
2.木制地板和天花板切合商店自然主题
3.设计精巧祥和的通道
4.俯视大厅
5.店内各个商品区融合自然
6.宽敞的空间能给顾客更好的品牌体验
7.体验、休息区

to take their clients to try on new products. The five projects gather the same spaces: product exhibit and trying area, two or three auditoriums, a coffee house, administration, a storage area and a sup.

The architectural and furniture design were worked together, so to achieve the development of a cohesive set, functional, with great esthetic appeal. Some icons (such as the entrance tunnel, wooden brise-soleil, water mirrors and indoors and outdoors gardens) were created so that, despite the characteristics of the purchased property, every Casa Natura displays the same features, recognisable by clients. Thereby, these first five houses are considered the embryo of an initiative to be extended all over the country.

1. Entrance 1.入口
2. Reception 2.接待处
3. Products 3.产品展示
4. Makeup area 4.化妆区

3

4

5

6

7

1

Healthy Spot

健康点

Location:
Santa Monica, USA

Designer:
Akar Studios

Photographer:
Randall Michelson

Completion date:
2008

项目地点：
美国 圣塔莫尼卡

设计师：
阿卡工作室

摄影师：
兰德尔・迈克逊

完成时间：
2008

Healthy Spot is set in the midst of high-end shops and restaurants on a trendy section of Santa Monica's Wilshire Boulevard. The 270-square-metre space is highlighted by a wide glass frontage that gives the specialty dog products store an open and inviting feeling. Consisting of a series of interconnected spaces: the retail showroom displaying a variety of accessories, a day-care space for dogs with its own "grassy" patch, and a grooming salon where customers can intermingle, the impact of the design concept for the space is immediately apparent from the modern composition of the interior layout and the surface finishes used.

Within the main retail display space which painted in white, the strategically placed fixtures define the circulation pattern through the primary section of the store. Arranged in a concentric formation, the display fixtures are set off from the floor to allow space underneath to be visible and open – in the process, providing a floating effect for the display of the merchandise. Furthermore, this layout allows unhampered movement of the customers and their dogs. For creating the accents and providing a colourful contrast to the white washed interiors, custom-designed colourful wall murals grace the main walls of the store.

健康点位于圣塔莫尼卡的威尔希尔大道上，周围林立着高端店铺和餐厅。门口开阔的玻璃墙面给这个宠物用品商店带来了开放而诱人的感觉。店铺由一系列的连通空间组成——展示配件的零售样品间、铺满人造草坪的日间宠物犬托管区、顾客可以参与的宠物美容沙龙。现代的室内布局和表面装饰凸显了项目的设计理念。

主要商品展示区被漆成了白色，固定装置的设置确定了店铺的流通区域。按照同心构造排列的展示装置，保证了下部空间的可见性和开放性，为陈列的商品提供了一种悬浮的效果。此外，这种布局还保证了顾客和宠物犬的无障碍移动。为了强调白色的室内空间，打造色彩对比，设计师特别在墙壁上添加了定做的彩色壁画。

2

1. An open and clear storefront
2. The display fixtures are set off from the floor to allow space underneath to be visible and open
3. The interior lighting is lively and bright
4. The display area

1.开放而通透的店面
2.展示柜与地面分离，保证了空间的可见性与开放性
3.室内灯光活泼明亮
4.商品展示区

1. Office 1.办公室
2. Retail display 2.零售区
3. Day care 3.日间护理区
4. Shampoo 4.洗护区
5. Grooming 5.修剪区
6. Restroom 6.洗手间

3

4

1

LeFel Store

莱菲尔家居馆

Location:
Milan, Italy

Designer:
Crea International

Photographer:
Crea International

Completion date:
2009

项目地点：
意大利 米兰

设计师：
克里国际设计公司

摄影师：
克里国际设计公司

完成时间：
2009

Crea International, a retail design company headquartered in Milan, has designed for LaFeltrinelli Group, the concept for the new brand LeFel. The leading idea was based on the theme of travelling through a "whispering world". A world that aims to whisper to people able to listen, leading them to reach for something new. Travelling in the LeFel world means to open heart and mind, put stories together, tales, emotions that people usually put back in their intimacy and sometimes pull out to warm soul and share with adventure companions. The new LeFel store intends to be a place for people's souls where a free and adventurous spirit, sometimes eccentric, puts the most significant objects of its wandering, travel by travel.

It is a place where words, sounds, scents, past and future meet and where the objects tell stories and talk to people who have chosen them. It is a mutable and essential place where mysterious boxes, as transparent and light as thoughts at times, or as sturdy as wood other times, pick together objects that whisper little, big stories to the people who are able to listen to them.

The new LeFel format wants to express the concept of entertainment and client faithfulness through a continuous renewal of the proposals that fit tendencies. In LeFel environment the graphic mark is strong. It was

克里国际设计公司为拉菲尔特瑞奈利集团策划设计了新品牌莱菲尔，其主要设计理念是在“细语世界”中旅行，让人们在轻声细语中获得全新的体验。在莱菲尔的世界中漫游意味着打开心胸，将内心深处温暖人心的故事与感动于同行者分享。莱菲尔家居馆让我们自由而热爱冒险的灵魂找到了栖息之所，在漫游中找寻自己所期待的物品。

这是一个汇集了文字、言语、气味、过去和未来的地方，商品不断地诉说着自己的故事。像是一个多变而神秘的盒子，时而如思想一样透明而轻盈，时而如树木一样坚定，人们会在这里找到能够与自己进行交流的适合自己的物品。

莱菲尔的新模式通过不断的潮流更新表现了消遣的理念和顾客的忠诚度。在莱菲尔家居馆中，图像符号异常突出，以达到让商品向消费者讲述故事的目的。旅行的理念被转换为图形，展现在商品的邮戳上，以见证这次愉快的感官之旅。

木材以一种温暖的方式包裹着整个环境，暖色的家具采用了橡木、胡桃木和黑铁木。中心区域层层叠叠的展台吸引了顾客的目光，而LCD屏幕则促进了店铺与顾客的交流。

WHAT'S N

HOME
ART
DESIGN

LEFEL

1. The display of various products
2. The bicycle seems to be telling its story
3. The space arrangement is delicate and careful
4. The gallery area on the first floor
5. The entrance area
6. Balloons with the store's logo on them

1.丰富多样的商品展示
2.店内摆设的自行车仿佛在讲述着自己的故事
3.空间布局精巧用心
4.二楼的画廊区
5.商店入口处
6.印着商店标志的气球

conceived to offer the client a journey inside the LeFel store: a shopping experience where objects tell about the people who choose them. The leading idea of the travel was then transferred graphically in the postmark as sign of something that can be brought home after a joyride that transmits strong sensations.

Wood and its warmth wrap the environment, using the space properly. For the furnishing, wood was used in its warmest tones: oak, walnut, wenge. The focus is in the central zone dedicated to new arrivals. A "lab tree", made up of modular overlapping elements, attracts the attention of clients also thanks to LCD screens that implement communication.

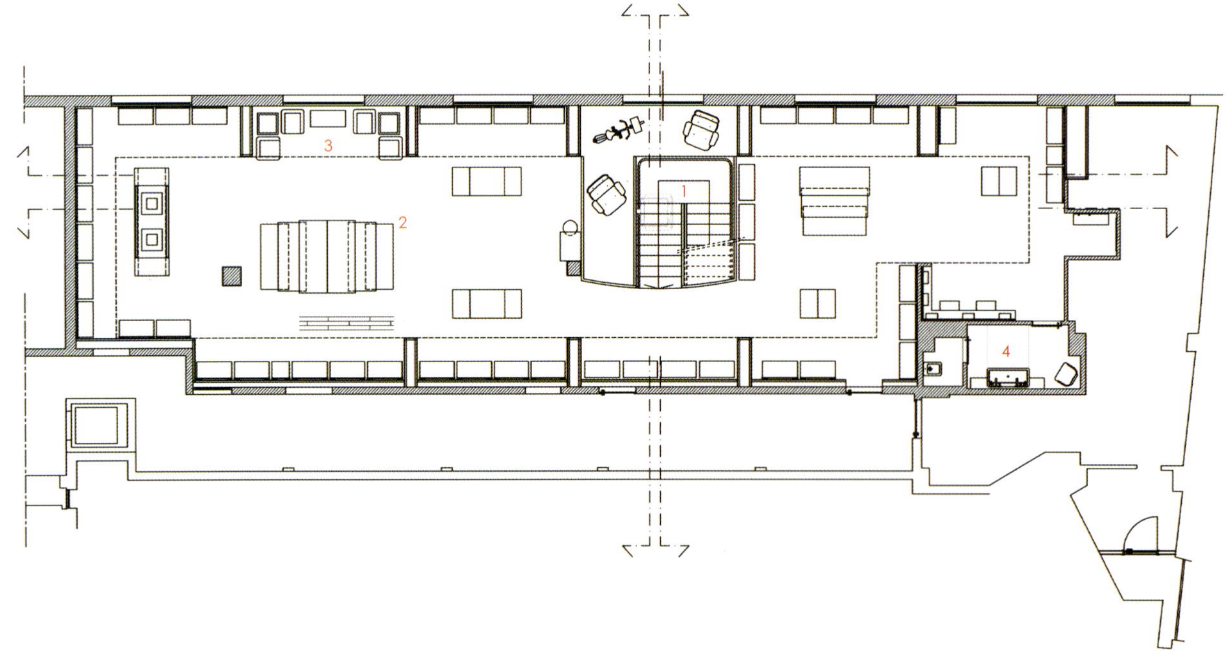

1. Stairs 1.楼梯
2. Display area 2.展示区
3. Rest space 3.休息区
4. Washroom 4.洗手间

3

4

ALLARME
INCENDIO
entrata

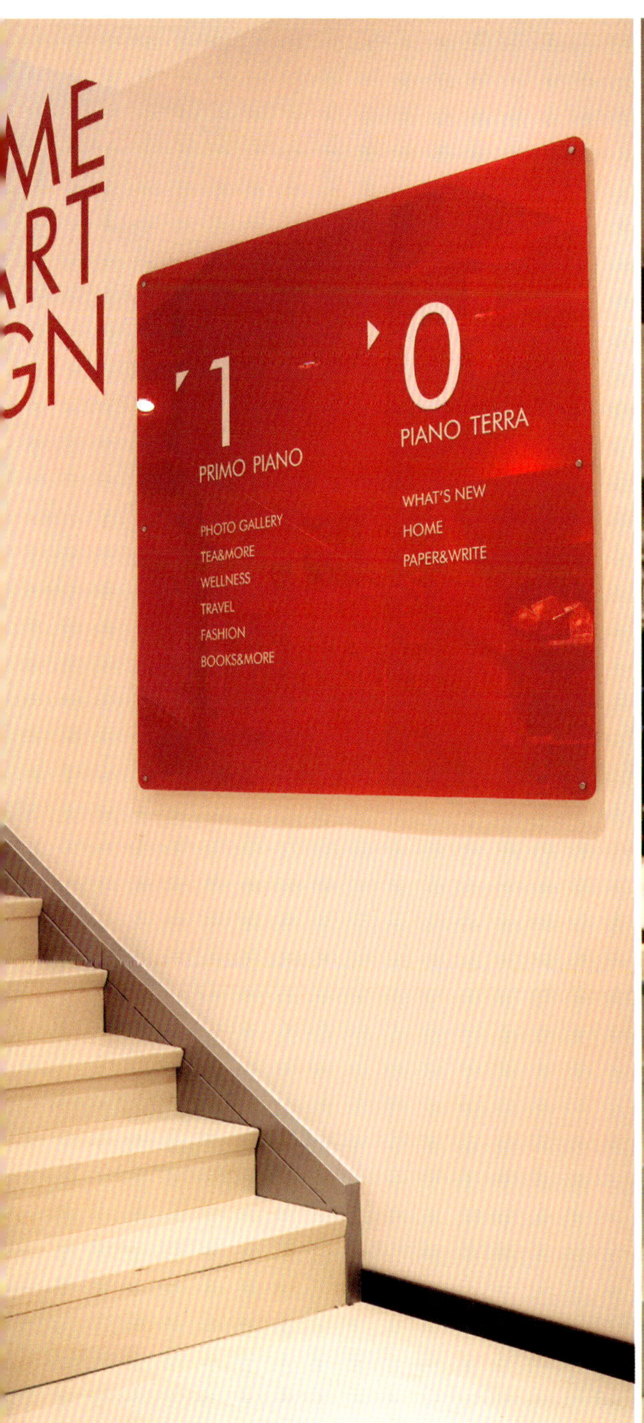

1
PRIMO PIANO
PHOTO GALLERY
TEA&MORE
WELLNESS
TRAVEL
FASHION
BOOKS&MORE
0
PIANO TERRA
WHAT'S NEW
HOME
PAPER&WRITE

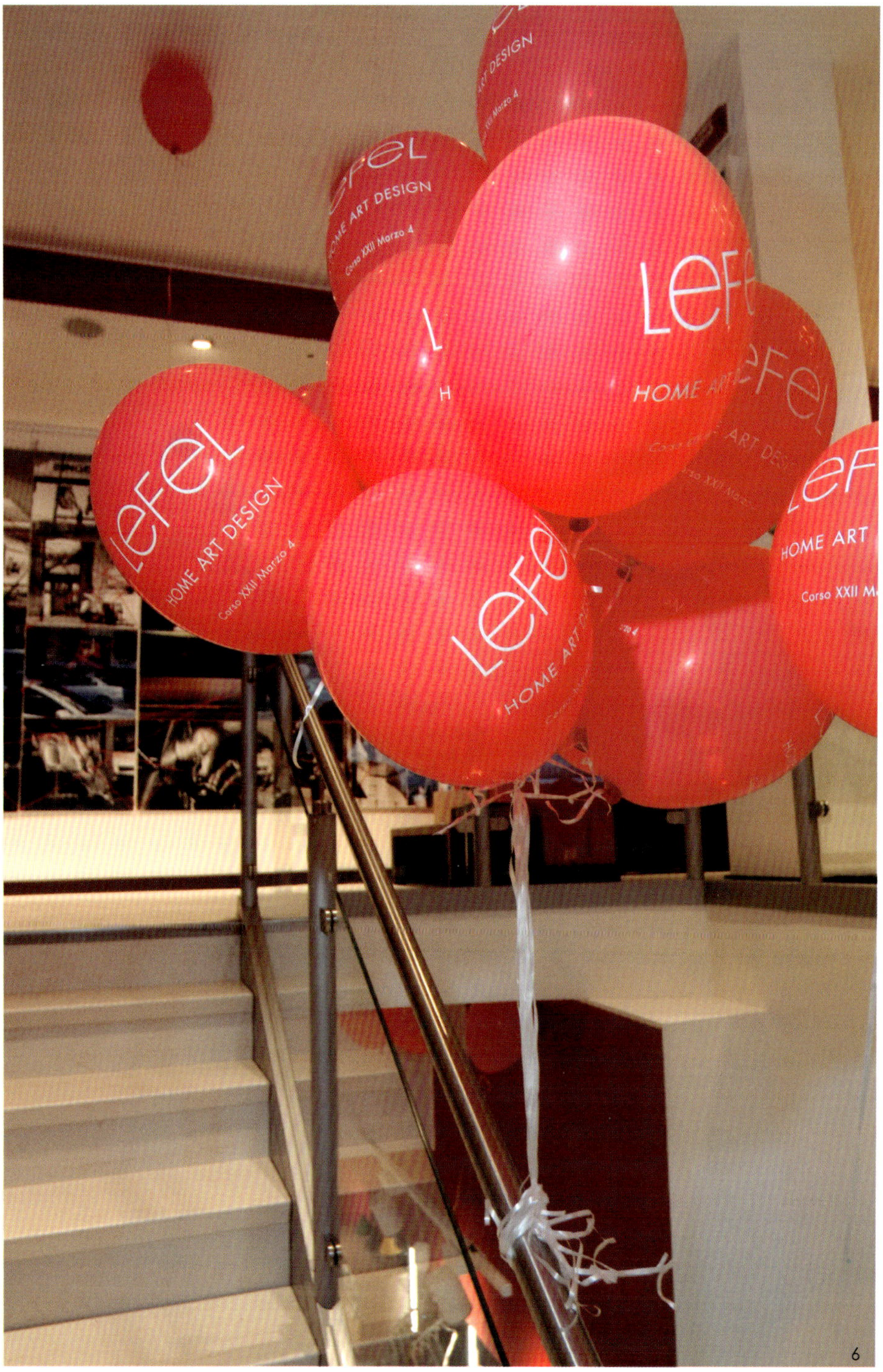
LEFEL
HOME ART DESIGN
Corso XXII Marzo 4

Legacy Books

遗产书店

Location:
Plano, USA

Designer:
Morrison Seifert Murphy

Photographer:
Morrison Seifert Murphy

Completion date:
2008

项目地点：
美国 普莱诺

设计师：
莫里森·赛福特·墨菲

摄影师：
莫里森·赛福特·墨菲

完成时间：
2008

Legacy Books is an independent bookstore located in Plano, Texas, an affluent suburb of Dallas. One might think it foolish to build a bookstore in the current climate of online book sellers and retail chains losing money. How does one draw the public into a bookstore when they can receive the exact same product online? The solution was to create a store that served as much as a public space (like a hotel lobby or department store) as it did a bookstore, with interlocking spaces and upper floor "galleries" to view the action within. With this in mind, the designers broke apart the original two-storey vacant shell space – eroding away the floor plate and sliding mezzanine floors into that void. Instead of losing space the designers kept the same amount of square footage, while increasing the perceived size of the interior. The plan includes interesting features such as the café, demonstration cooking area, a Wi-Fi bar, event space, and the art & architecture library.

The plan is organised around the primary design element: the elevator. Each floor floats from it and stairs wrap around it. Pressed aluminium panels surround the elevator to create a skin of typeface letters, reminiscent of printing presses. Also unique to this bookstore is the centrally located demonstration cooking area which allows chefs to show off their cooking skills while promoting their latest books. The display cooking area can also be viewed from multiple

遗产书店位于达拉斯市郊的普莱诺，是一家独立书店。在网络售书和零售连锁书店纷纷亏本之际，人们或许认为建造一座书店是件愚蠢的事情。怎样才能把消费者从网上书店拉回来？遗产书店的解决方式是让书店成为一个公共活动空间，如酒店大堂或是百货商店。设计师打破了原来的双层空壳结构，移除了楼板和滑动阁楼楼板，形成了一个开阔的空间。尽管如此，设计师还是通过增大室内感知空间保证了总建筑面积没有缩减。项目中设置个形形色色的特色区域，如咖啡厅、烹饪展示区、无线上网吧、活动空间和艺术与建筑图书馆。

项目的设计围绕着基本设计元素——电梯展开。每层楼都漂浮在电梯之上，而楼梯则环绕着它们。电梯周围的铝板上被压印上了铅字字体，令人想起了报业印刷。书店中心的烹饪展示区也是一绝，大厨们可以边展示厨艺边宣传自己的新书，各个楼层都能看到烹饪展示区。每个不同的图书部门都设置着休闲座椅区，优雅的书架和特别定制的家具让遗产书店与传统式书店截然不同。

2

1. The bookstore has a spacious space
2. The generous staircase
3. The novel and beautiful shelves
4. The space has a bright and warm colour palette
5. The store provides customers with various services
6. The white shelves highlight the colourful books

1.书店内空间十分宽敞
2.开阔的楼梯
3.书架造型新颖美观
4.店内色调鲜艳温馨
5.商店内部设施功能多种多样
6.白色书架凸显了书籍的缤纷

levels in the bookstore. The layout features a variety of casual seating areas within the different book sections with elegant shelving and other custom fixtures, all unlike what is found in those typical bookstores.

1. Entrance
2. Café
3. Terrace
4. Demonstration kitchen
5. Front of house
6. Periodicals & gifts
7. Fiction
8. Children's
9. Sort room
10. Break room
11. Office
12. Seating

1.入口
2.咖啡厅
3.露台
4.展示厨房
5.外区展示
6.期刊和礼品
7.小说区
8.儿童图书区
9.分类室
10.休息室
11.办公室
12.休息区

Level Two
EAT
SLEEP
READ
BargainBooks
BargainBooks
BargainBooks
Level One
3

Holiday
Cheerios
NUTCRACKER
Glittery Critters
Silent Night
jingle bell baby!
DRUMMER BOY

4

188 - 189

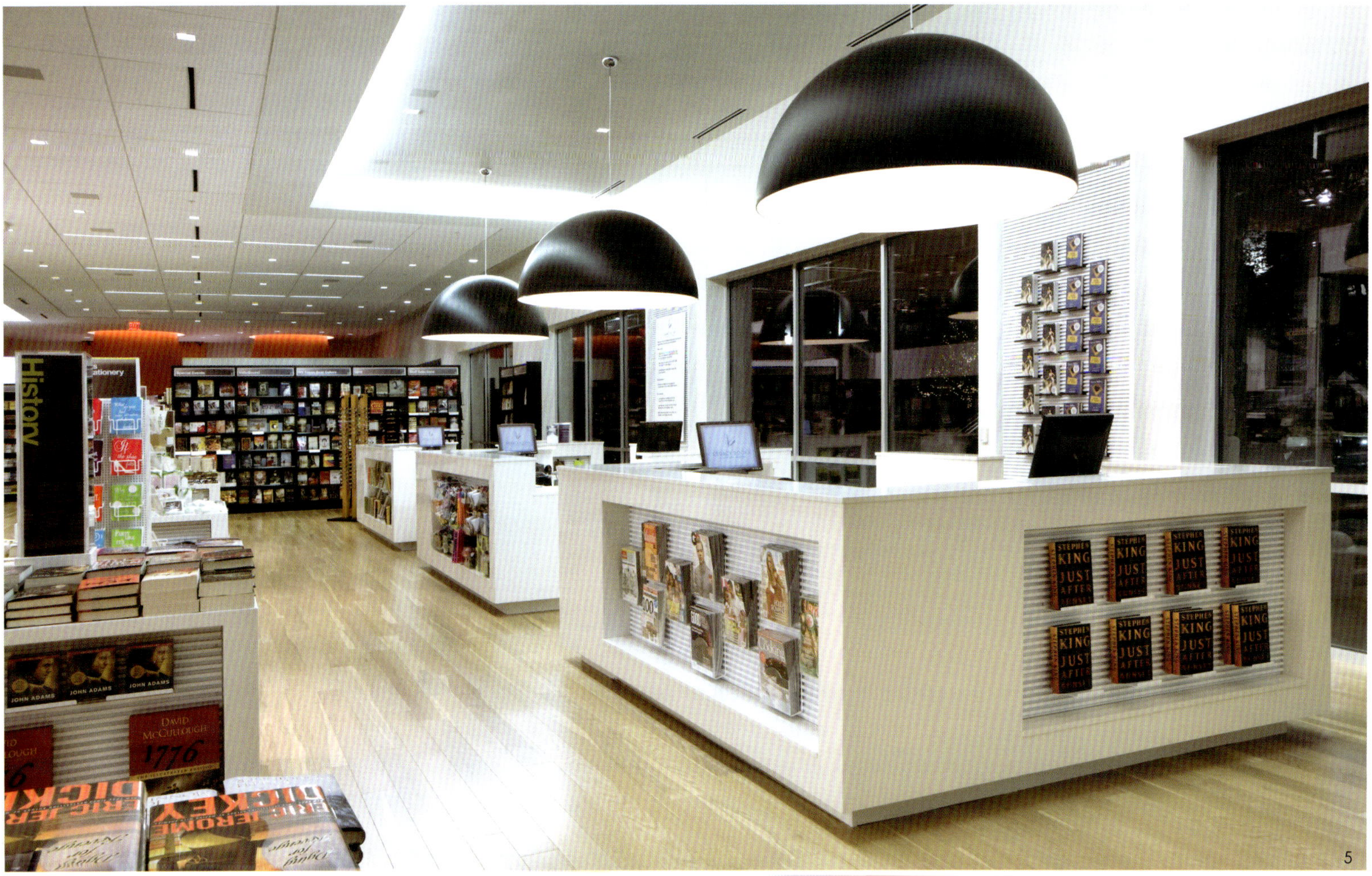
History
Stationery
STEPHEN KING
JUST AFTER SUNSET
JOHN ADAMS
DAVID McCULLOUGH
1776
5

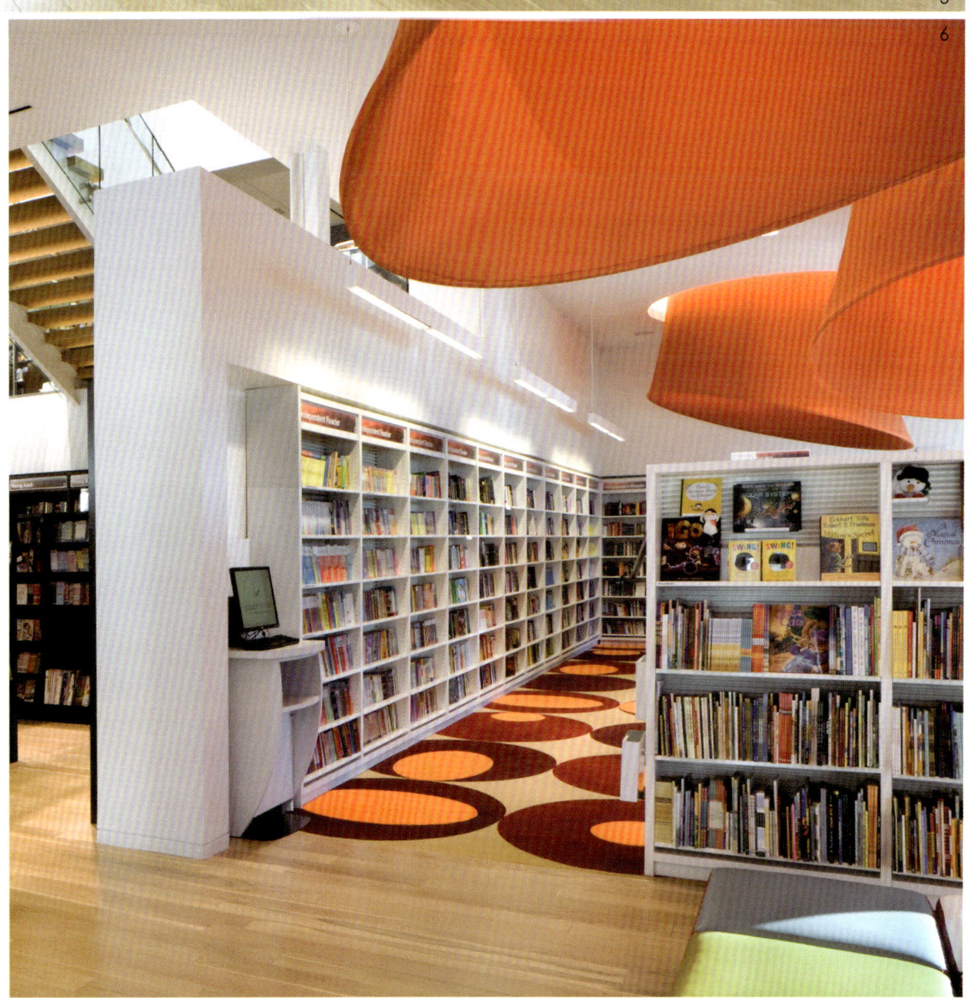

6

London Luxury

伦敦精品

Location:
New York, USA

Designer:
D-ash Design

Photographer:
Brad Dickson

Completion date:
2008

项目地点：
美国 纽约

设计师：
d字灰设计公司

摄影师：
布莱德·迪克森

完成时间：
2008

London Luxury, a wholesaler of fine bath towels and accessories, asked D-ash Design to design a new showroom in New York City. The showroom was to be a retail interpretation of a luxury spa. The concept was to create a sophisticated yet quiet environment and allow the product to shine. The focal material utilised in the space was a sustainable wood floor made of bamboo, which was installed on floors and walls. The product was chosen for its durability and sustainability but also for its close appearance to teak. The display system for the product was all custom designed out of industrial components, which complemented the "rough" appearance of the loft showroom.

伦敦精品是一家高档浴巾和浴室配套用品专营店，d字灰设计公司受雇为他们在纽设计一家新的展示厅。展示厅被设计成一个奢华的水疗浴室，环境优雅而宁静，衬托出产品的高端品质。空间设计所采用的主要材料是铺设在地面和墙壁上的可持续竹子板材。之所以选择竹子作为材料，是因为它兼具耐用性和可持续性，并且在外观上类似柚木。产品的陈列架和展示台全部是特别定制的工业组件，彰显了展厅的粗放特质。

ALLERGY LUXE
ALLERGY LUXE

1. It is a peaceful and elegant space
2. The chandelier and products express a luxury style
3. The towel and bathrobe area
4. The interior decoration has a natural atmosphere
5. The main material used in space design is bamboo
6. A corner of the display area

1.店内设计优雅宁静
2.吊灯与商品展现了奢华的风格
3.毛巾浴袍销售区
4.拥有自然气息的室内装饰
5.空间设计使用的主要材料是竹子板材
6.商品展示一角

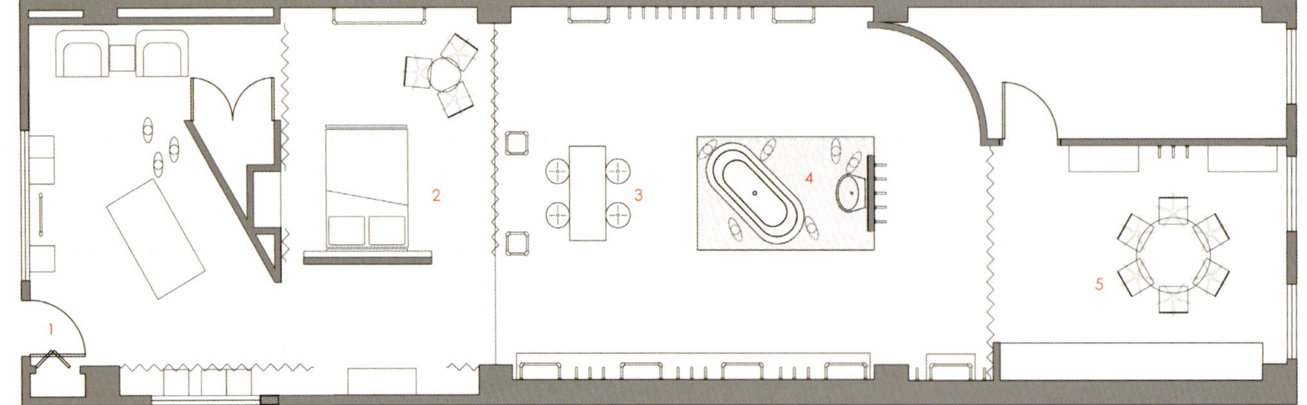

1. Entrance 1.入口
2. Bed 2.床
3. Display area 3.展示区
4. Tub 4.浴缸
5. Office 5.办公室

4

5

1

Royal Tea Barn

英格茶叶店

Location:
Beijing, China

Designer:
BMA Beijing Matsubara and Architects

Photographer:
He SHU

Completion date:
2008

项目地点：
中国 北京

设计师：
BMA 北京松原弘典建筑设计咨询有限公司

摄影师：
舒赫

完成时间：
2008

This project is an interior renovation for a tea shop located in Sanlitun Village, the new commercial city blocks in the central district of Beijing City. In addition to designing the east building on the north block of the village, BMA also designed interior of this tea shop on the south block. This store is in the building that Kengo Kuma, a Japanese architect, designed.

Despite of the small interior space with 63 square metres, the client has requested BMA to design various functions of spaces, including the meeting space for customers, the tea tasting space, and the product exhibition space. Therefore, one of the biggest themes for this project was to transform this small space to the one with the impression of the larger room, but at the same time designing it for multiple functional spaces. For those reasons, the designer has come up with the method, called "scene of a vision caused by creating the curved-surface wall". The three-dimensionally curved wall exists in the centre of the store space. The wall is split in three pieces so that a view penetrates through the splits to the various directions. For example, when people stand at the entrance of the store, they can view the cashier through the slit between the walls. When people stand in front of a wall on the east side, they can view a wall on the west side through the slit. Viewing an object through the slit between the walls, rather than viewing the exposed object without looking through

此次由BMA进行内装设计的茶叶店，地处北京市中心街区的新型商业街“三里屯village”。可巧的是，BMA之前亦参与了“三里屯village”项目的设计工作，其间担任的是北区东栋的建筑设计。而BMA此次主持内装设计的茶叶店位于“三里屯village”项目南区、日本建筑师隈研吾氏设计的建筑体中。

甲方希望在总计63平方米的狭小空间中兼顾茶品展示、茶品试饮、茶事商谈等多个功能区。由此，如何将狭小的空间进行划分实现多种功能并存，同时又能有效降低狭小空间带给人的压迫感，成为了该项目设计上的重要课题。针对这一点，设计师在设计中，采用了“借用曲面墙重新打造视觉空间”的设计方法。在这个狭窄的内部空间中，他们在中央部分设计了三维立体弯曲墙面，该墙面自然过渡，分为三段，重塑了室内的视觉空间。站在入口处，透过曲面墙可以看到收银台。站在东侧墙面前，亦可以透过曲面墙看到西侧墙面。对面墙体上的茶品若隐若现，无形中增加了空间的纵深感，从而在一定程度上实现了将狭小空间适度扩展的可能。高约3.5米的天花，与狭长走向的空间布局结合在一起，也增加了一定的开放感。

另外，中央部分的曲面墙端部与东西两壁相连。东西两壁与此构成了二维曲面，东西两壁开槽，作为茶品的展示台。西侧作为试饮区，依弧形墙面设沙发，墙端回旋设洗手池。曲面墙表面全部着乳白色，墙面镂空及开槽处着以粉红色，强调曲面。其他墙面及家具全部上红色，与曲面墙形成对比。在曲面墙的掩映下红色家具若隐若现，增加了空间的纵深感，从而从一定角度实现了视觉空间的扩展。

2

1. The entrance
2. The curving wall has an exquisite design
3. With the white wall as a backdrop, the semi-circle sofa is highlighted
4. The red furniture increases the horizontal depth of space
5. The curving wall improves the interior visual effect
6. The display of the products

1.店面入口
2.弯曲的墙面设计十分精巧
3.白墙映衬下的环状沙发十分醒目
4.红色家具若隐若现，增加了空间的纵深感
5.曲面墙增强了店内的视觉效果
6.产品展示

any slit, creates an impression of the depth in the space. Because the ceiling is very tall as 3.7 metres, the interior space becomes long, vertical, and open space.

Moreover, the end of central wall is spread and connected to the east and west walls. The east and the west walls also have two-dimensional curved surfaces. The shelves are fixed inside the slits of these walls and tea leaves are exhibited in the shelves. On the west side of the store, sofas for tea tasting zone are placed along the wall and create circular form. The end of curved-surface wall on the west raps around a small kitchen for the tea tasting. Although the entire wall surface was painted in white, the slits and the bonders of the walls were all painted in pink to emphasise the curved-surface of the walls. The individual furniture detached from all the walls is in red to present the drastic contrast to the rest of the walls. These hidden and sought views of the red furniture emphasise the spatial depth of the store.

1. Tea-making area 1.煮茶区
2. Porcelain display 2.瓷器展区
3. Tea display 3.散茶展区
4. VIP area 4.贵宾区
5. Cashier 5.收银台
6. Brand display 6.品牌展区
7. High quality tea 7.精品展区
8. New arrival 8.新品展区

4

The Best of Asia is Makaibari.
Royal T- Barn
Royal T
Royal T-Barn
Peppermint
Royal T-Barn
Hibiscus
Royal T-Barn

5

6

1

Sensora Day Spa

桑索拉日间SPA馆

Location:
Sydney, Australia

Designer:
Design Clarity

Photographer:
Design Clarity

Completion date:
2009

项目地点：
澳大利亚 悉尼

设计师：
明晰设计公司

摄影师：
明晰设计公司

完成时间：
2009

At Sensora Day Spa there is a focus on service, indulgence and ultimate luxury. With a series of individual beauty treatment rooms, discreet "microzones" for fast-turnaround mini facials, a "skinbar" for product dabbling and makeovers, a comfortable communal hairstyling area and bespoke retail product display zones, Sensora offers the complete package. The store environment is sleek and minimal, restrained and sensual – answering the brief for the creation of an evocative space for sensory retailing.

Inspired by the science of skincare, Design Clarity created a distinct brand mark for Sensora reflecting the molecular makeup of skin cells. This logo device has been manipulated into a fragmented pattern and used as a recurring motif throughout the interior in blonde timber screening devices, sculptural wall pieces, mirrored panels, three-dimensional ceiling planes and subtle light features. The built environment and graphics have a consistent thread, which continues through the custom-made packaging concepts, uniform design, ticketing and even product shelf talkers. The shop front design extends this scattered fractal pattern to the store façade across the frameless glass, inviting and intriguing customers to enter under a portal of pure white vitrified ceramic mosaic tiles.

桑索拉日间SPA馆以服务、放松和终极奢华为中心。桑索拉提供全套的美容服务，包括独立美容室、快速面部美容区、产品试用和化妆的美容台、舒适的发型设计区和定制零售商品摆放区。店铺环境简约而时尚、矜持而奢华，为商品提供了一个灵动的空间。

明晰设计公司受到护肤科学的启发，为桑索拉设计了一个独特的商标，商标反映出了皮肤细胞的分子组成。这个标识被分割成碎片图案，应用在金黄色木屏风上、墙壁浮雕上、镜板上、3D天花板上和精细的灯光装置上，贯穿了整个室内空间。建筑环境和图形设计贯穿一线，延续了统一的定制包装理念、制服设计、标签和货架插卡。店面设计进一步将分散的碎片结构扩展到了无框玻璃墙上，贴满纯白陶瓷锦砖的大门迎接着四面八方的顾客。

室内装饰的选择反映了自然而简约的美学观：美国橡木地板、白色皮革、镜子、低挥发性涂料和纯白色细木家具。白墙上的白色装饰和亚光标牌上的光泽装饰为室内增添了精妙的质感，强化了桑索拉的品牌形象。日间SPA馆占据了两间店铺的空间，并倾斜45度角，形成了一个动感的空间。展示货架和展示墙正对进入店铺的顾客。

1. The entrance with pure white tiles
2. The molecular skin cells motif extends to the glass wall
3. It is a fashionable and clean design, low-key but luxurious
4. A stand for testing products and make-up

1.贴满纯白陶瓷锦砖的入口
2.商店的分子状标识延续到玻璃墙上
3.店内设计时尚简约，矜持而又奢华
4.产品试用和化妆的美容台

Finishes have been selected to reflect a natural minimal aesthetic: American Oak flooring, white leather, mirror, low VOC paints, and white solid surface joinery. White on white walls and gloss on matt reflective signage elements add another layer of subtle texture to the interior and reinforce the Sensora brand. The day spa layout covers two adjacent retail tenancies and is set out at a 45-degree angle creating a more dynamic sense of space and allowing the display units and half height merchandise walls to face the customer on approach.

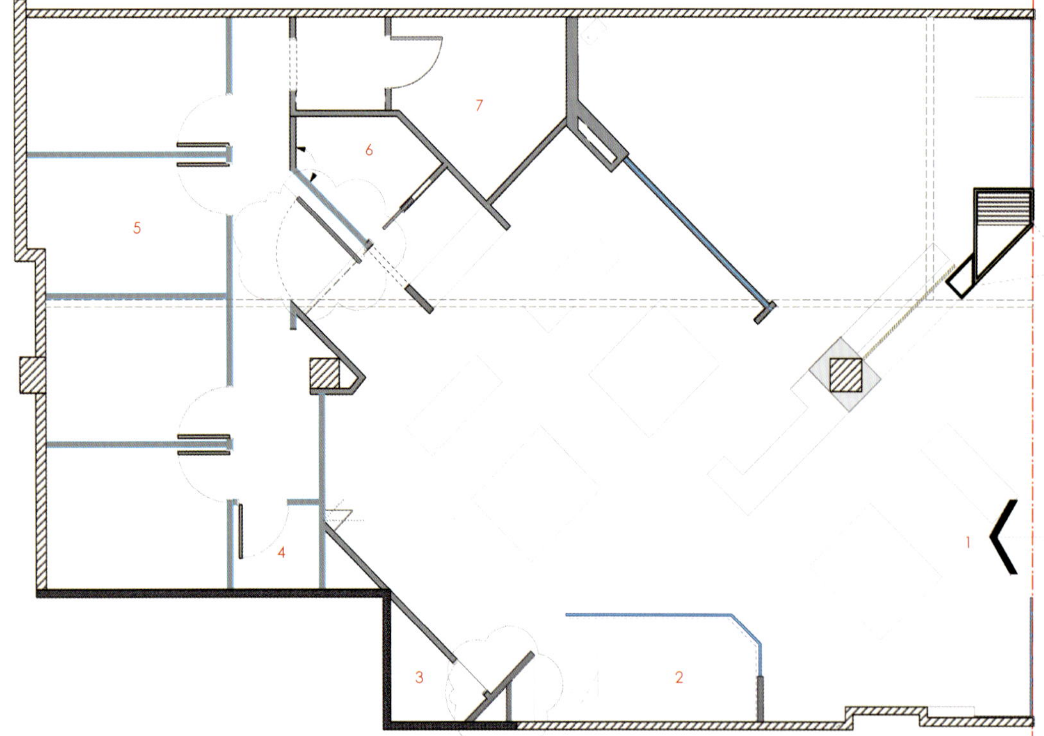

1. Entrance
2. Microzone room
3. Store room
4. Change room
5. Treatment room
6. Managers room/store room
7. Kitchenette/staff

1.入口
2.快速美容区
3.储藏室
4.更衣室
5.处置室
6.经理室/储藏室
7.厨房/员工休息室

3

4

1

Store at the Museum of Arts and Design

艺术设计博物馆商店

Location:
New York, USA

Designer:
JGA

Photographer:
Laszlo Regos Photography

Completion date:
2008

项目地点：
美国 纽约

设计师：
JGA设计公司

摄影师：
拉罗兹·里格斯摄影公司

完成时间：
2008

With a focus on the implicit and subtle transformation of materials, this retail environment is a natural place for consumers to look for objects from serious to whimsical, functional to esoteric, and complex to simple. Positioned off the main lobby at the heart of New York's Westside cultural district, the Store becomes the perfect setting to showcase the Museum's unique assortment of saleable art and merchandise. The space is wrapped on three sides by the building's new glass façade. Iconic "lollipop" columns exposed from the building's structural legacy provide pattern and mass to the space with both a shadowy exterior presence and strong geometric interior rhythm and form. Key elements of the museum store include an elliptical curved glass showcase, ceiling mounted merchandising "trees", and the main focal element: a forty-foot-long organically-curved translucent glass merchandising installation.

The main focal element of the shop is a forty-foot-long by eight-foot-high organically-curved glass installation. Through its translucency, the nature of the glass provides a hint of the colourful glass and art merchandise through shadowing to the street, particularly in evening hours. During the day, the internal lighting is augmented by natural daylight, heightening the colour and translucency of the objects from an interior perspective in a way that provides both a neutral and complementary

商店设计中材料的运用十分精妙，是一个让顾客可以随意挑选不同种类、不同功能物品的场所。商店位于纽约西部的文化街区，是博物馆销售艺术品和商品的最佳场所。店铺的三面是建筑的新玻璃幕墙。建筑标志性的“棒棒糖”支柱为空间提供了形象的图案，让室外空间变得朦胧，也让室内空间充满了几何韵律和造型。博物馆商店主要设计元素为椭圆形玻璃展柜、吊顶树枝以及12米长的玻璃展架。

商店的焦点元素是12米长、2.4米高的弧形玻璃展架。半透明的展架透过街道的阴影衬托出彩色玻璃工艺品，尤其在晚上更为明显。白天，自然光线增强了内置灯光装置的效果，强化了商品的色彩和透明度，为这些独一无二的艺术品提供了自然而现代的背景。这个展架是商店的标志性元素。

椭圆形玻璃展柜主要展示手工制作的珠宝首饰，共分为三层，内置LED照明装置。椭圆形的周长近13米，中心的功能岛用于储藏商品、提供检修服务和进行焦点陈列。展柜由可丽耐大理石和星火玻璃制成，商品放置在特别定制的分层展台之上。

2

1. A general view of the store
2. The oval glass showcase has a beautiful profile
3. There are various products in the store
4. The display of handmade jewellery
5. The white streamline display shelf fills the space with rhythms

1.店铺纵览
2.椭圆形玻璃展柜造型优美
3.店内商品种类繁多
4.手工制的珠宝首饰展示
5.流线型的白色展架让空间充满韵律

background to these one-of-a-kind pieces. It becomes an iconic and highly identifiable signature object to the store visitor.

An elliptical curved glass showcase features handcrafted and one-of-a-kind wearable art (primarily jewellery) in a three-tier, internally LED-illuminated presentation. The forty-two-foot circumference of the ellipse features a centre functional island for storage of additional pieces, a service point and an opportunity for focal displays. The case is faced in Corian and Starfire glass, and products are uniquely merchandised on compressed felt custom-configured and layered pads.

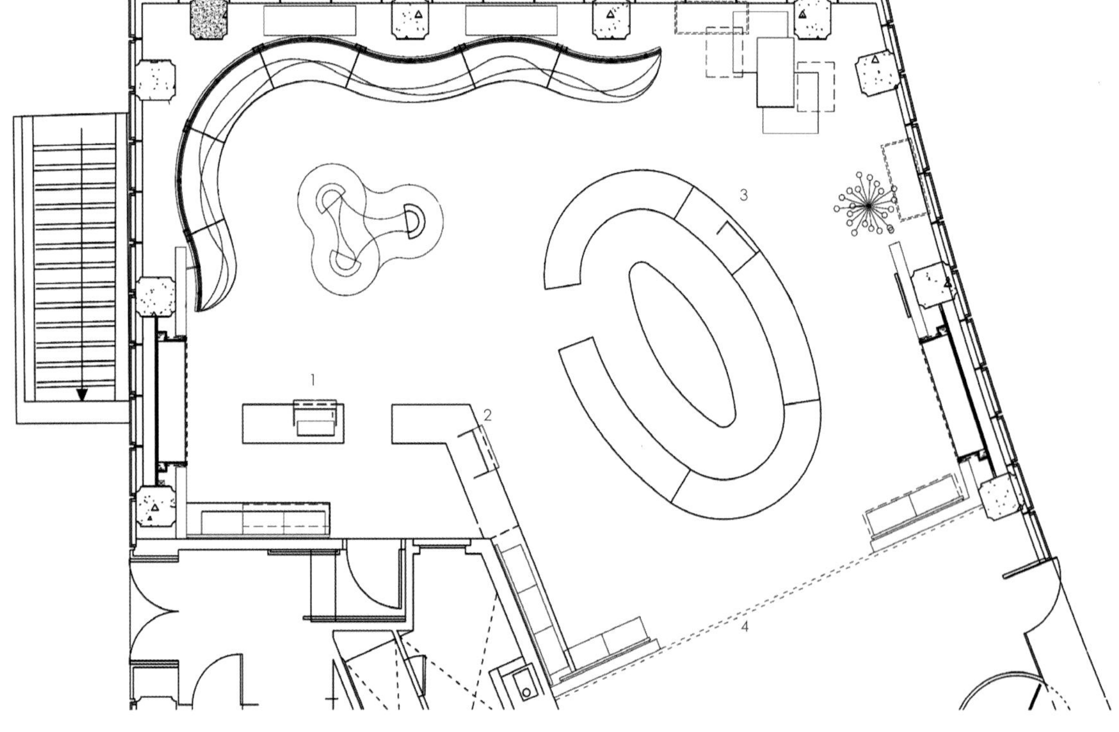

1. Register 1.登记处
2. Counter 2.柜台
3. Glass counter 3.玻璃柜台
4. Store front 4.商店正面

3

4

1

A1 Shop

A1旗舰店

Location:
Vienna, Austria

Designer:
BEHF Architects

Photographer:
Bruno Klomfar

Completion date:
2007

项目地点：
奥地利 维也纳

设计师：
BEHF建筑事务所

摄影师：
布鲁诺·克洛姆法

完成时间：
2007

In the course of the rollout of an estimated forty-three A1 shops, BEHF Architects have architecturally defined the brand value of the company in the clear and simple language of forms. The individual elements of this shop concept are highly flexible in terms of multi-use possibilities. This flexibility permits a uniform brand image despite very different spatial and functional circumstances.

Mobilkom Austria with its Premium brand A1 is a company not only successful for its brand values, including personal touch, innovation, service and quality; its slogan "We'll connect what connects you" also comes alive architecturally in its new shops.

Open spaces do not allow any barriers between customers and sales staff, and the counter placed in the middle of the room (meander) plays a pivotal role. It serves on the one hand as a point of first contact; on the other hand, the customer is invited to move through the whole room to explore the A1 product worlds on his own.

该项目是第43家A1旗舰店，BEHF设计师用清晰简明的设计理念将该公司的品牌价值诠释出来。该设计理念中的个性元素灵活地体现于商店的不同功能和区域之中。尽管各空间及功能不尽相同，但这种灵活性仍然服务于品牌形象的统一性。

奥地利移动业务提供商Mobilkom Austria及其经典品牌A1店的成功不仅在于它给人个性的观感、巧妙的创新、优质的服务和过硬的质量，同时还有它的广告标语——“你需要的就是我追求的”，这些都使得该新店的设计充满个性。

开放的空间让顾客和售货员之间零距离接触，柜台设在大厅正中间，是整个商店的中心枢纽。这样设计一方面便于与顾客进行沟通；另一方面便于引导顾客找到他们需要的A1产品。

2

1. The exterior view of the shop
2. A display stand in unique form
3. The space design is open and free
4. A small meeting room behind the curtain
5. Customers can choose their favourite products in a comfortable environment
6. The relaxing area

1.商店外观
2.造型独特的展示台
3.空间设计开放自由
4.幕帘后的小型会议室
5.顾客可以舒适的在体验中选择自己喜爱的产品
6.店铺休息区

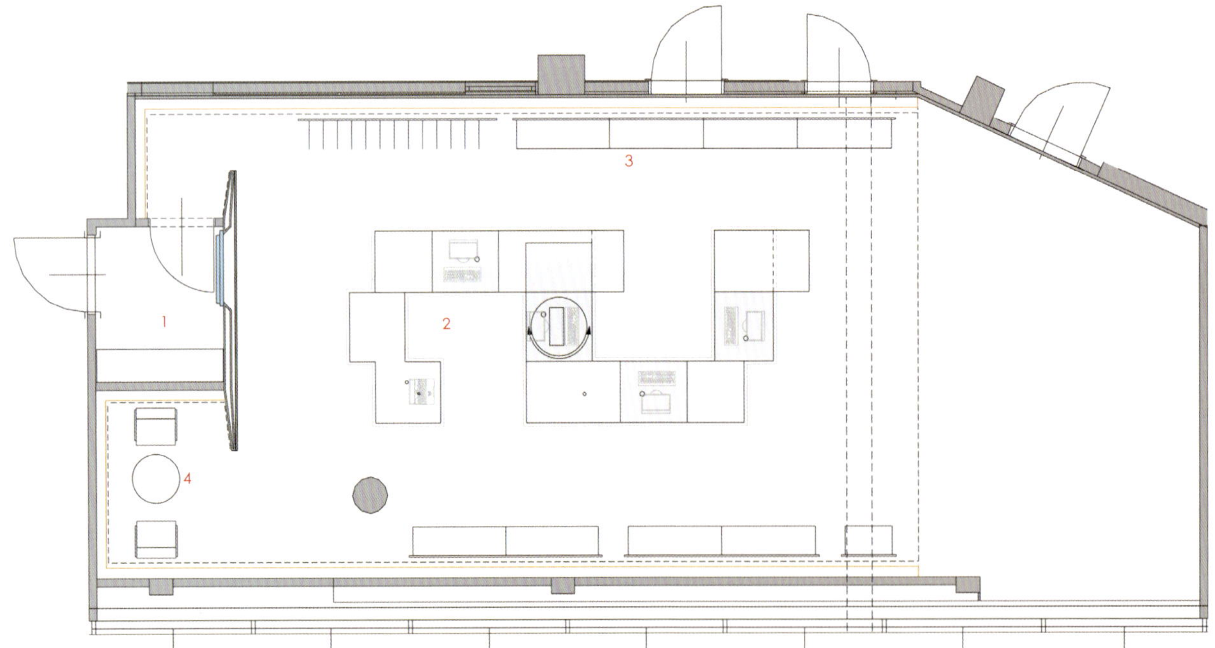

1. Entrance 1.入口
2. Counter 2.柜台
3. Display area 3.展示区
4. Rest area 4.休息区

3

4

5

6

BASE Flagship Store

BASE旗舰店

Location:
Antwerp, Belgium

Designer:
Creneau International

Photographer:
Philippe Van Gelooven

Completion date:
2010

项目地点：
比利时 安特卫普

设计师：
克莱诺国际设计公司

摄影师：
菲利普・凡・格鲁温

完成时间：
2010

The central station of Antwerp is the location of the BASE Flagship Store. The building is a classified monument with an enormous history. It makes a very solid impression. On the other hand, the BASE brand values are: user-friendliness, clarity, transparency and emotion. To merge these two strong identities in one location, the designers of Creneau International played with contrasts; the massive historical building with its heavy materials against the light, fresh and clear elements of the BASE brand.

The BASE brand identities lead the designers to creating a concept that refers to a greenhouse: a place where costumers find themselves in a high-tech environment with a very simple and low-tech feel but always transparent and clear. An environment that breathes "BASE" in every way possible. With respect to the monumental building, Creneau International treated the whole shop as a "box in a box". All integrated elements are standing free of the walls. In this way, the ancient building becomes a historical backdrop, a decor in which BASE is the leading player.

BASE is hot! Which is why the designers created a super stylish greenhouse for the newest flagship store of the Belgian mobile operator. From now on, the brand is growing and greening in the warmest corner of the majestic Central Station of Antwerp. Also very cool: the airy and transparent contrast of the hothouse with the massiveness of the historical station.

BASE旗舰店位于安特卫普的中心车站，其所在的大楼是一座经典历史建筑，令人印象深刻。另一方面，BASE的品牌理念是：方便用户、清晰、透明、真情实感。为了将这两个鲜明的特性聚集在同一地点，克莱诺国际设计公司的设计师采用了对比设计：历史建筑厚重的材料与BASE品牌新鲜、轻盈而清晰的品牌理念形成了强烈的对比。

设计师根据BASE的品牌特征设计了一个“温室”。顾客可以在一个简单而不具技术含量的透明空间里尽享高科技产品环境。这个环境的每个角落都散发出BASE品牌的气息。为了像历史建筑致敬，设计师把店铺设计成了“盒中盒”，所有的装饰展示元素全部远离了墙壁。这样一来，古建筑形成了空间的历史背景，是BASE旗舰店的装饰。

BASE品牌新潮而惹火，因此设计师打造了一个极其时髦的温室环境。这样一来，BASE品牌就可以在宏伟的安特卫普中心车站最温暖的角落茁壮成长。此外，轻快透明的温室也和厚重的车站建筑形成了鲜明对比。

2

1. The old building and the novel brand make a sharp contrast
2.The interior design is light and exquisite
3. Customers can enjoy the high-tech products better in a comfortable environment
4. Print-on-demand: only specific folders for clients are printed
5. The store expresses a novel and flexible design style

1.古老的建筑与新颖的品牌标识形成鲜明对比
2.店内设计精巧轻盈
3.开放的空间能让顾客更好的感受高科技产品
4.按照需求来打印：代替一整本宣传册，顾客可以在此只打印他们需要的文件信息
5.新颖灵活的设计风格得到很好的体现

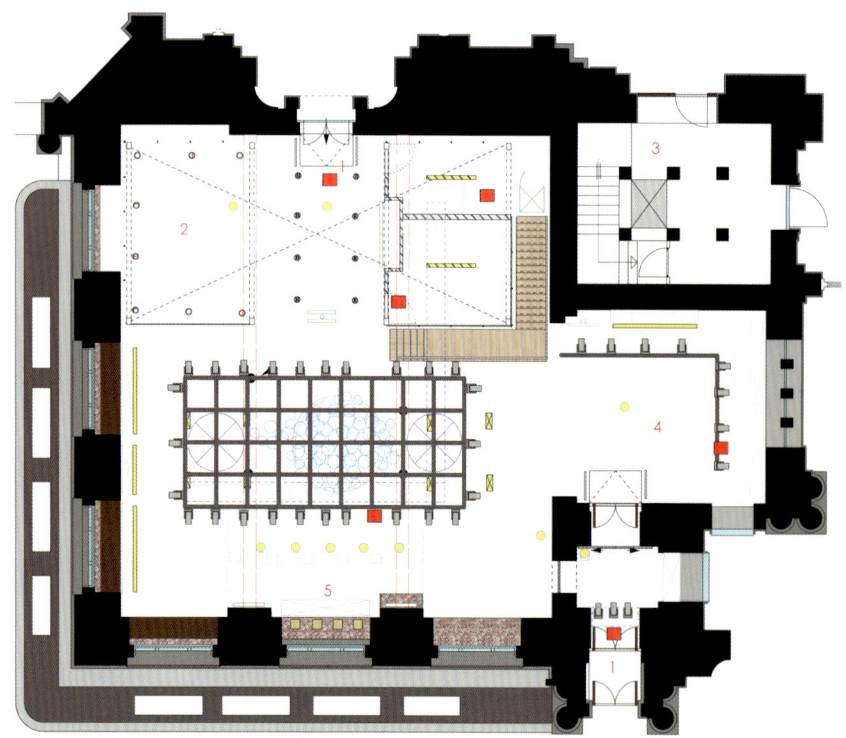

1. Entrance
2. New product presentation
3. Back office
4. Printing zone
5. Counters

1.入口
2.新品展示
3.后部办公室
4.打印区
5.柜台

SAMSUNG
NOKIA
freedom of speech
info info info info
ACTIVEER NU
RIHANNA
RELOAD = DOWNLOAD
HERE

BASE is ook internet voor thuis en onderweg
BASE
freedom of speech
105
204
313
427
BASE
jim mobile
BLACK EYED PEAS
RIHANNA

4

smartphones
BlackBerry
freedom of speech
try me
freedom of speech
NOKIA

1

FlatFlat in Harajuku

平平体验馆原宿店

Location:
Tokyo, Japan

Designer:
SAKO Architects

Photographer:
Chikao TODOROKI/Sasaki Studio Inc.

Completion date:
2008

项目地点：
日本 东京

设计师：
迫庆建筑事务所

摄影师：
等等力周夫/佐佐木工作室

完成时间：
2008

FlatFlat in Harajuku is the store where the visitors are actually able to experience Hangame, an online games portal, and Hange.jp, a games portal for mobile phones, developed by NHN Japan Corporation. The store is located at the centre of Harajuku, a part of Tokyo, known internationally for its youth style and fashion. The store occupies a narrow space of 3.5 metres wide by 45 metres long with complicated plan and structure as the result of repeated reconstruction.

The concept of the store is the future park and the designers attempted to combine virtual element with real space open to anyone. It is not only the space where "organic principle" and "inorganic principle" stay together, but it represents the modern society which consists of real and virtual environments.

"Organic principle" consists of curved lines which have a space characteristic of a cave where among the various spaces people can discover their own place to stay depending on each purpose and feeling while the different scenes unfold continually. The designers attempted to create a space where people feel like snuggling up to the organic form that curves based on human body dimensions. Visitors do not hesitate to enter the store because of the sense of closeness. On the other hand, "inorganic principle" consists of white wall surface fixtures, neon tubes of ceiling illumination and mortar

平平体验馆原宿店是一家可以让顾客真实体验Hangame在线游戏门户网站和Hange.jp手机门户网站的体验中心。体验馆位于原宿（东京以潮流时尚而著称的区域）的中心，长45米，宽3.5米，是一个狭长的空间，内部设计采用了复杂的重复结构。

体验馆的设计理念是打造一座未来公园，设计师试图将虚拟和现实融合在一起。这个空间不仅将有机元素和无机元素结合在了一起，而且是现实和虚拟相融合的现代社会的缩影。

有机元素由曲线组成，体现了山洞的特性，人们可以在自己的空间里，实现不同的目标，体验不同的场景。设计师想打造一个让人们可以依偎的有机结构，因此，弧形结构的设计都是以人体尺寸设计的。空间的私密感让人能放心进入。另一方面，无机元素由白墙、天花板上的霓虹灯管和水泥地板组成，营造出虚拟感。设计师运用霓虹灯管和地面接缝的线条来映衬墙面装置的形状。它们激起了顾客的好奇心，将顾客们引入狭小空间的内部。

1. The layered streamlines
2. The curving structures make the store look like a future park
3. Customers can enjoy a virtual on-line game
4. Reality and virtuality combine well in the store

1.重叠的流线造型
2.曲线结构仿佛一座未来公园
3.顾客可以真实体验在线游戏
4.现实与虚拟在店内得到很好的融合

floor creating virtual character. The designers used the lines which let neon tubes and the prevention of crack seam of mortar offset the forms of the wall surface fixtures. They stimulate curiosity of the visitors by synergy with the forms of the fixtures and lead them to the inner part of the narrow space.

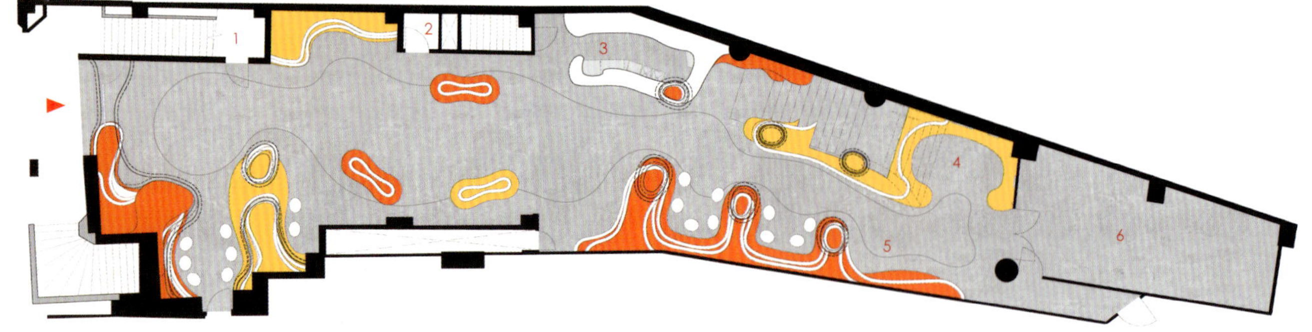

1. Ware house　1.仓库
2. Eps　2.应急电源
3. Kitchen　3.厨房
4. Sampling section　4.样品区
5. Promotion area　5.促销区
6. Office　6.办公室

3

4

1

O2

O2体验店

Location:
London, UK

Designer:
JHP

Photographer:
JHP

Completion date:
2008

项目地点：
英国 伦敦

设计师：
JHP设计咨询公司

摄影师：
JHP设计咨询公司

完成时间：
2008

Design consultancy JHP has been appointed by O2 to work on the design of a new store concept for the telecoms retailer. The store environment seamlessly links the worlds of the contemporary home and future technology without alienating the everyday customer. Contemporary concrete floor tiles and dark timber flooring are used to define different zones within the store. Overhead, the unique ceiling features sinuous interlocking blades, reflecting the form of sound waves.

At the back of the store, is the "lounge" area. Relaxed and inviting furniture positioned around a coffee table faces a huge interactive touch-screen which will show product demonstrations, gaming, video clips from O2 and from sponsored sports events. Timber flooring, textured wall-covering and lampshades bring warmth to the environment and adds to the homely feel. The lounge area will also provide a space for O2's presentations of "It's Your Community" where grants are presented to local community groups.

The white unitary on the shop-floor has been designed to flex as the product range evolves. Customer experience in-store will be enhanced further with the introduction of a self service table enabling customers to check their account, top up, and buy O2 event tickets. The planning of the store has been carefully built around the needs of O2's

电信设备零售商O2公司委托JHP设计咨询公司为其进行新店的设计。店铺环境将现代家居和未来技术紧密地结合在一起，并没有疏忽了日常的顾客。风格现代的地板砖和深色木地板分别代表了店内的不同区域。天花板上装饰着独特的蜿蜒交错的刃片，反映了声波的形态。

店铺后部是休息区，悠闲而舒适的家具围绕着咖啡桌摆放，正对着一个互动触屏。触屏可以进行产品演示、游戏和播放视频剪辑。木地板、纹理墙壁板和灯罩为室内环境带来了暖意和居家感。休息区还提供了O2公司的“这是你的社区”展览，他们为本地社区群体提供资金援助。

店铺的白色地板可以根据产品系列灵活调整。自助桌台的引入进一步提升了顾客的店内体验，顾客可以自助查询自己的账户、充值和购买O2活动券。店铺的平面设计围绕着顾客的需求精心设计，让他们可以更方便的选购配件、电话和其他电信设备。广告公司VCCP所打造的视觉营销与产品紧密相连，方便了店内的方向定位。

新的设计理念具有“绿色”特征，采用了节能照明、数码售票（替代了纸质票据）、数字显示屏（替代了海报）和调节型电源（可以根据不同环境需求调节）。

Accessories
iPhone
iPhone
Self Service
Phones
Every photo's a keeper
Business
Take the web with you
LG Mobile
We're better, connected

1. The display of products
2. Self-service area
3. Customers can experience the high-tech products
4. The unique ceiling features sinuous interlocking blades, reflecting the form of sound waves
5. The comfortable relaxing area
6. The clear and deep entrance

1.商品展示台
2.自助服务区
3.顾客可以体验高科技产品
4.天花板蜿蜒交错的刃片，反映了声波的形态
5.舒适的休息区
6.通透深邃的店铺入口设计

customers who wanted an easier way to shop for accessories, phones and other telecoms devices. Advertising agency VCCP has created visual merchandising that connects directly to the product and category facilitating ease of orientation around the store.

The new concept also boasts "green" credentials employing the use of energy efficient lighting, digital ticketing instead of paper, digital screens instead of posters and managed power sources which accommodates different day part requirements.

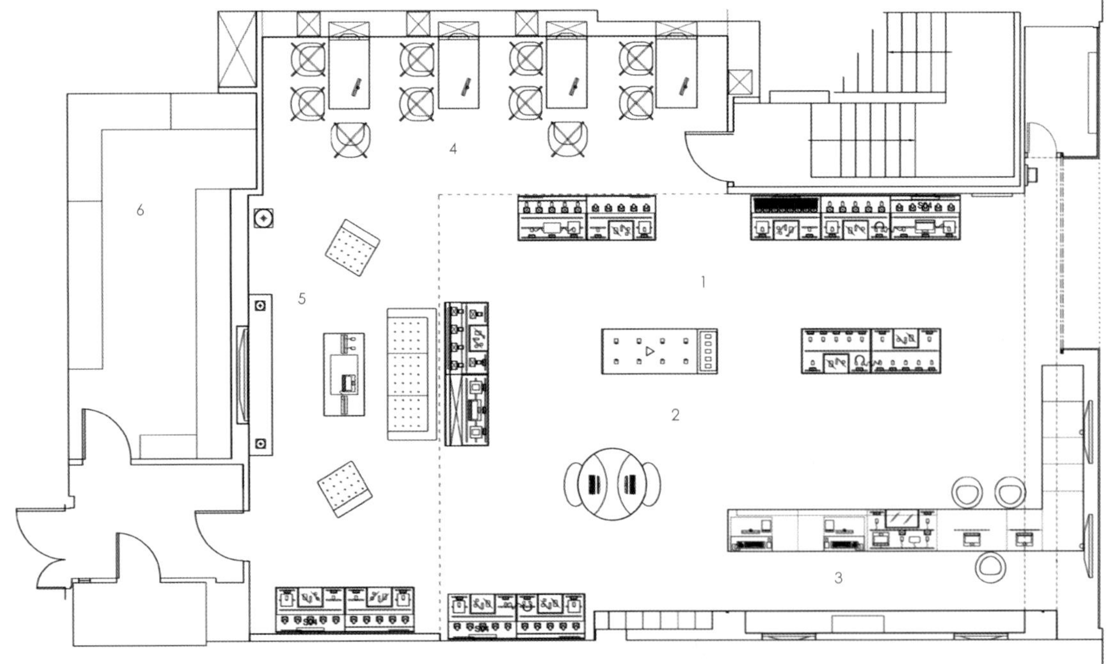

1. Browse
2. Retail area
3. Guru bar
4. Membership sign up
5. Live product play
6. Store area

1.浏览区
2.零售区
3.吧台
4.会员签约处
5.现场产品试用
6.储藏区

Sony Ericsson
Sony Ericsson
Sony Ericsson
Bluetooth®
speakers
on C902

We Recommend
Accessories
Broadband
Sim Cards
Phones
4

5

Sony Ericsson
W910i
£99.99

PRIORITY

6

1

One2free Flagship Store

自由2旗舰店

Location:
Hong Kong, China

Designer:
Curiosity, Gwenael Nicolas

Photographer:
Curiosity

Completion date:
2008

项目地点：
中国 香港

设计师：
Curiosity设计公司，格温尼尔・尼古拉斯

摄影师：
Curiosity设计公司

完成时间：
2008

One2free is the number one mobile brand in Hong Kong. The new interior is a materialisation of the graphic identity. The elements of the logos become three-dimensional displays. The consistency in the communication is an important factor for a successful brand. The message is unique and can be easily comprehend by the customers.

The elements of the logo, the circle is applied in the interior as well as the façade. The shop exterior becomes an iconic expression of the brand identity. In the entrance a large "digital chandelier" created with circular elements integrate screens and visuals. Reflected on the ceiling the circles appear as the drops of a "digital fall".

Clarity, transparency, fluidity reflect the brand essence, and also become inspiration for the interior space, displays and lighting. Unique interior design systems were developed to offer new experiences to the customers. The new handsets are displayed in large water drop displays, and they appear floating in the middle of the space. Dual screens integrated on counter top, offer unobstructed interaction between the staff and the customers. The shop offers different zones depending on the type of services, on the top floor a digital centre, framed in an orange glass arena offering individual services. The iconic shape and colour become a strong symbol of the new interior identity.

自由2是香港排名第一的移动通信品牌。自由2旗舰店的室内设计是图形标志的融合。品牌标识被制成3D造型，进行展示。沟通的一致性是成功品牌的重要因素，品牌所传达的独特信息需要做到通俗易懂。

品牌标识的圆形造型被应用在室内设计和外墙设计中，店铺的外观形象地表达了品牌形象。入口处巨大的数码吊灯融合了屏幕和视觉造型，从天花板上垂下来，圆形元素宛如数码瀑布的水滴。

自由2的品牌核心是清晰、透明、流畅，也是室内设计、展示设计和照明设计的灵感来源。独特的室内设计为顾客提供了全新的购物体验，新款手机被展示在大型水滴造型展区，仿佛悬浮在空中。案台上的双面屏幕为工作人员和顾客提供了无障碍互动。不同的服务被分配在不同的区域，顶楼的数码中心四周环绕着橙色玻璃罩，提供个人服务。自由2品牌的标志性形象和颜色成为了室内设计的重要元素。

品牌的橙色元素被应用在不同区域，以引导顾客，与白色的地板、天花板和家具形成了鲜明对比。手机被展示在黑色亚光玻璃橱窗内，宛如黑色屏幕上的像素点，反映了通信工具的数码特征。

1. There is a large digital screen at the entrance
2. The store design features a sense of future
3. The circles appear as the drops of a "digital fall"
4. The matt black display window
5. The space design is pure and unique
6. The interactive consulting area

1.入口处悬挂着巨大的数码屏幕
2.店内设计极具未来感
3.圆形元素宛如数码瀑布的水滴
4.黑色亚光玻璃橱窗
5.空间设计纯净另类
6.产品咨询互动区

The orange brand colour was widely used in the space to highlight the different zones and guide the customers. It is contrasted with the white of the floor, ceiling and furniture. The handsets are displayed on matt black large windows, so the customers can focus on them as pixels on a black screen; it expresses the digital and immateriality of the communication devices.

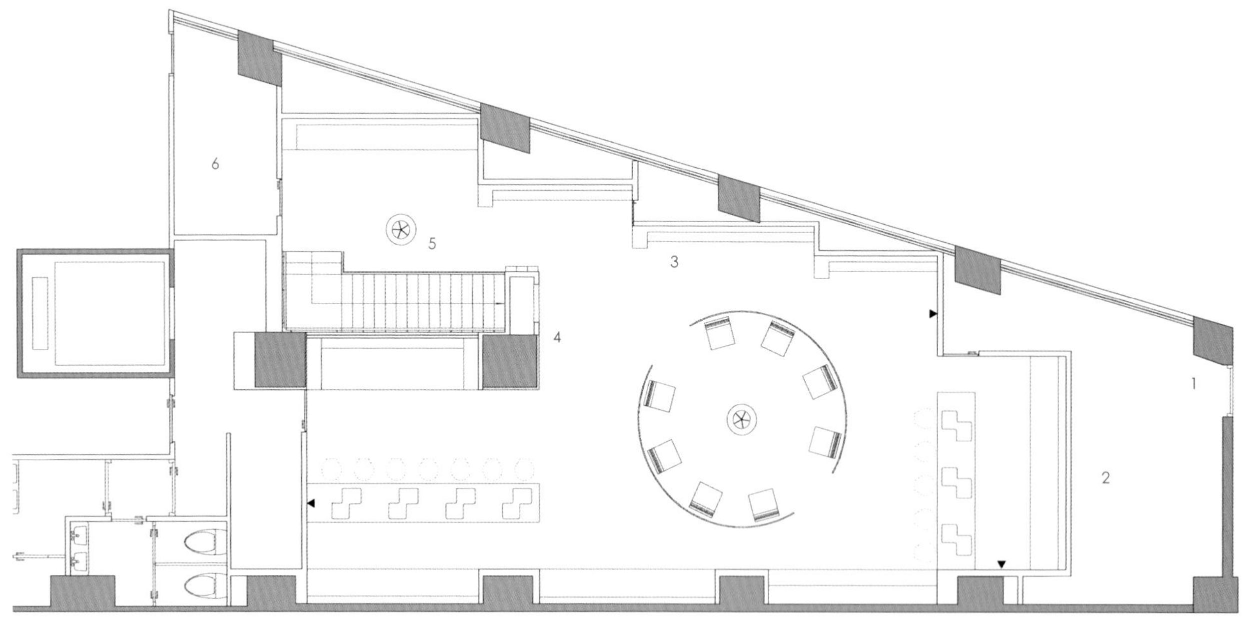

1. Exit
2. Office & stock
3. Display
4. Monitor
5. Handrail
6. Stock

1.出口
2.办公室&储藏室
3.展示区
4.监控器
5.楼梯扶手
6.储藏室

one2free
one2free

4

238 - 239

5

6

1

Tsunami Concept Store

海啸新理念商店

Location:
Oporto, Portugal

Designer:
Pedra Silva Arquitecto

Completion date:
2007

项目地点：
葡萄牙 波尔图

设计师：
pedra silva建筑事务所

完成时间：
2007

In the 19th century, when the dooes opended to reveal the new Tsunami Concept Store what initially seemed like an impossible mission was made possible. The expertise of the construction company BEC, was able to put a space of this complexity together in record time. This store, just as its contents used technology to make this space possible, from high-tech fiberglass panelling to CNC cut furniture. Opening was accompanied by celebrities and the presence of the press.

With this proposal, the designers were able to relate several aspects of quality and above all innovation. They sought to destroy preconceptions and developed this unconventional solution. The project team, composed by designers and architects from three nationalities, using hi-end software, were able to achieve an excellent result in spite of the short time frame available – and example of the ablity to deliver quality projects, with great challenges and tight deadlines.

19世纪，那些实干家开始展示起初看起来不可能完成的新的"海啸新理念商店"，而现在商店的建成已经证明了它的可能性。建筑公司工程合作委员会的专家意见能够将空间复杂性集合在一起。商店运用了高科技纤维玻璃板和电脑数值控制的切割设备等技术。在开张时有大量名人和媒体到场。

根据这个建议，设计师能够联系到一些质量方面的问题，最重要的是改革与创新。他们寻求毁坏预想和发展这个非传统的解决的方法。项目研究小组，由三个不同国家的设计师和建筑师组成，采用高端软件，能够在短时间内完成这个商店，同时也证明了他们能够带着巨大挑战与短期限这样的压力来递送品质项目的能力。

2

1. The display area with rectangle niches
2. The ring is the centre of the store
3. The raised wall plate creates a mysterious future atmosphere with blue light
4. A general view of the store
5. The open interior space is convenient for customers to experience the products

1.矩形壁龛展示区
2.环状构造为店铺中心区域
3.光洁凸起的墙面板块搭配蓝光营造了神秘的未来氛围
4.店内纵览
5.开放的室内空间十分方便顾客体验产品

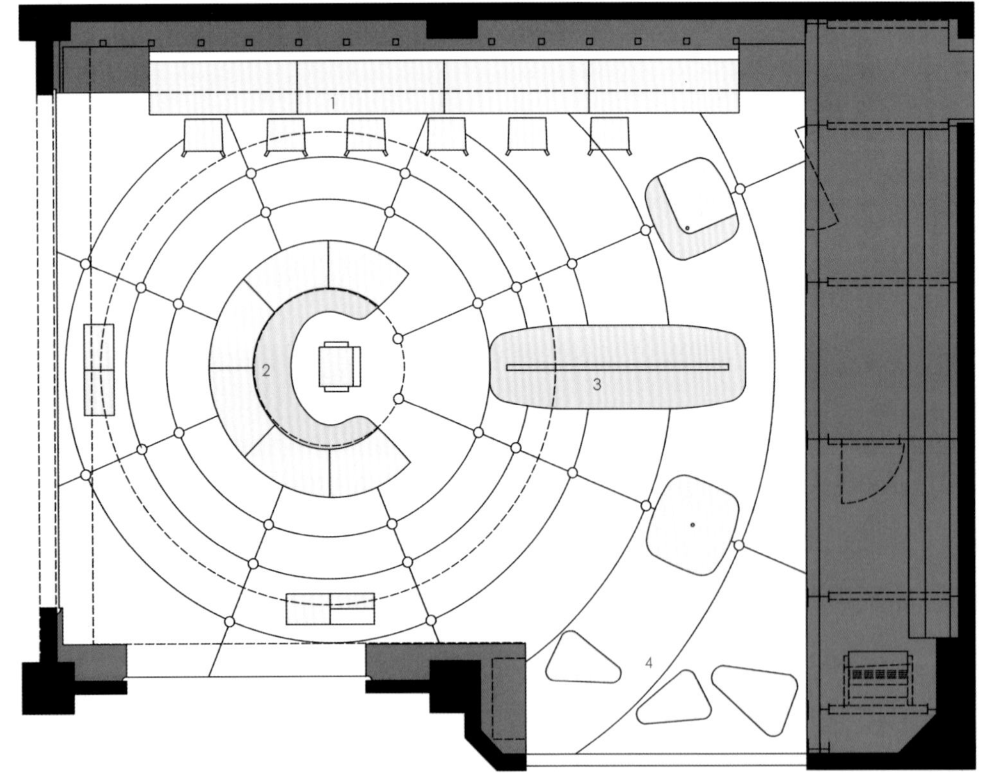

1. Display
2. Ring counter
3. Showing table
4. Rest area

1.展示台
2.环状柜台
3.展出区
4.休息区

TSUNAMI

4

TSUNAMI

1

Wind Retail Operative Platform

大风零售平台

Location:
Milan, Italy

Designer:
Crea International

Photographer:
Barbara Bozzi

Completion date:
2007

项目地点：
意大利 米兰

设计师：
克里国际设计公司

摄影师：
芭芭拉・博兹

完成时间：
2007

Crea International was asked to conceive for the Wind Retail Operative Platform to guarantee and keep constant communication of the brand regarding each single service it provides, through a clear, lively and colloquial tone.

The key word permeating the whole design concept is "Fashion Tech", which leads to a right stylistic mixture between a wiggly and soft design with the latest smart technologies, making the space synonymous of brand essence and values. So it is seen as a new means of communication and it has been chosen as the set for the Wind TV spots.

The environmental branding is mobile and particularly characterised by enlarged pixel which runs across the entire perimeter of the store and is the symbolic frame which contains visuals, signage for the services supply and product displays. Together with the orange outline, it punctuates the environment and represents the dynamic icon that generates the rhythm of the stores and develops their memorable language. The concept is wholly modular and it can be declined into different layout formats, but keeping the same effectiveness in terms of visibility.

克里国际设计公司为大风零售平台进行了构思，通过清晰、生动而通俗的设计将大风品牌的服务传递给顾客。

贯穿整体设计的关键词是时尚技术，这实现了弯曲、柔和的设计与最新智能技术的完美混搭，让空间与品牌精神和价值实现一致。这是一种全新的交流方式，该店面的设计已经成为大风品牌的电视广告背景。

设计的品牌化以移动通信为主，放大的像素块随处可见。作为具有象征性的框架，它们被用作视觉交流、引导标识和产品陈列。带有橙色轮廓的像素块装点着环境，形成了动态的图标，在大风品牌店营造了韵律感。整个设计都是模块化的，可以在不同的布局和层次中保持同等的视觉效力。

2

1. The reception of the store
2. The bold and novel design of the orange frame
3. The display windows are delicate
4. The whole interior space is decorated with orange pixels
5. The curving and soft experiencing stand design

1.商店接待台
2.橙色框架设计大胆新颖
3.造型精巧的橱窗设计
4.橙色的像素块装点着室内空间
5.弯曲柔和的体验台设计

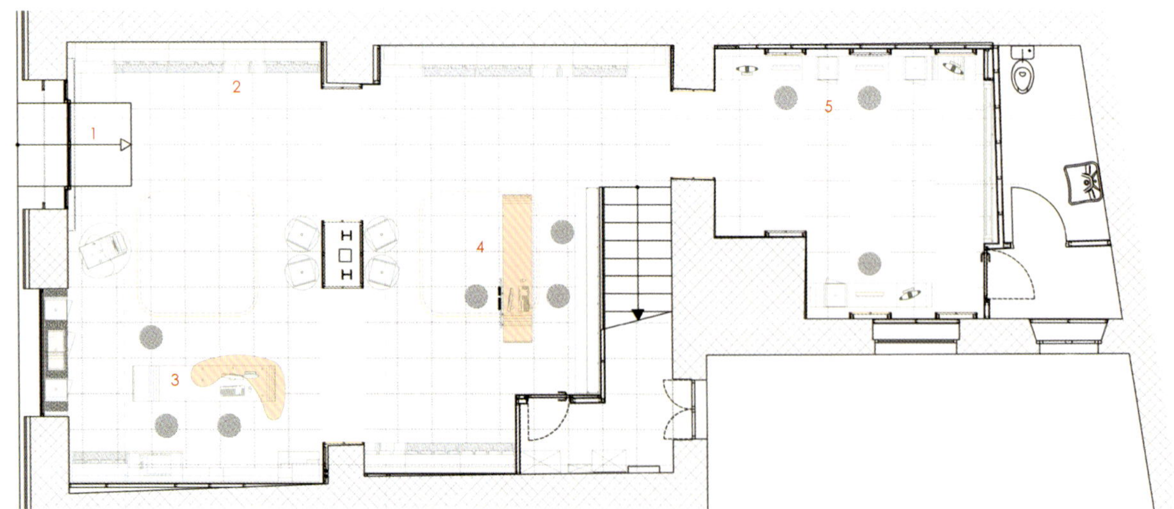

1. Entrance 1.入口
2. Display 2.展示区
3. Counter 3.柜台
4. Cashier 4.收银台
5. Experience area 5.体验区

WIND
INFOSTRADA
PROVA WIND senza cambiare
il tuo numero di telefonino.
PUOI PARLARE

4

INFOSTRADA
ADSL
4 Mega.
LIBERO

Barbie Shanghai

芭比上海旗舰店

Location:
Shanghai, China

Designer:
Slade Architecture

Photographer:
Iwan Baan

Completion date:
2009

项目地点：
中国 上海

设计师：
斯莱德建筑事务所

摄影师：
伊万·巴恩

完成时间：
2009

The first Barbie Flagship for Mattel holds the world's largest and most comprehensive collection of Barbie dolls and licensed Barbie products, as well as a range of services and activities for Barbie fans and their families. Mattel wanted a store where "Barbie is hero"; expressing Barbie as a global lifestyle brand by building on the brand's historical link to fashion. The designers were charged with the design of everything. Barbie Shanghai is the first fully realised expression of this broader vision. It is a sleek, fun, unapologetically feminine interpretation of Barbie: past, present and future.

For the new façade, the designers combined references to product packaging, decorative arts, fashion and architectural iconography to create a modern identity for the store, expressing Barbie's cutting-edge fashion sense and history. The façade is made of two layers: molded, translucent polycarbonate interior panels and flat exterior glass panels printed with a whimsical lattice frit pattern. The two layers reinforce each other visually and interact dynamically through reflection, shadow and distortion.

Upon entry, visitors are enveloped by curvaceous, pearlescent surfaces of the lobby, leading to a pink escalator tube that takes them from the bustle of the street, to the double-height main floor. The central feature

美泰公司的第一家芭比旗舰店拥有世界最大、最多的芭比娃娃系列藏品，并且能为芭比迷和他们的家人提供一系列的服务和活动。美泰公司希望店内的芭比是主角，并通过将品牌的历史价值和时尚结合起来，让芭比成为一种全球化的生活方式品牌。设计师必须要进行全盘设计，而芭比上海旗舰店则是第一家实现了以上构想的店铺。它是时尚、有趣的芭比形象的化身，融合了过去、现在和未来。

设计师在外墙设计上参考了产品包装、装饰艺术、时尚和建筑肖像，为店铺打造了现代化的形象，表达了芭比的前沿时尚感和历史。外墙分为两层：塑形半透明内层树脂板和印有古怪格子图案的外层玻璃嵌板。两层外墙相互作用，通过反射、阴影和扭曲增加了彼此的视觉效果。

一进入店内，顾客便被大厅优美的线条和光芒四射的表面装饰所包围了，粉红色的自动扶梯带顾客远离喧嚣的街道，进入芭比旗舰店的主楼层。店铺的中心是一个三层楼高的螺旋楼梯，楼梯两侧被800个芭比娃娃所包围，整个店铺都围绕着楼梯展开。

楼梯连接了三层零售空间：女人空间，主营女士时尚用品、服装、化妆品和配饰。玩偶空间，主营玩偶、玩偶配饰和书籍，也进行玩偶展示，内部包含芭比设计中心，女孩们可以在这里设计自己的芭比。女孩空间，主营少女时装、鞋子和配饰，内部包含芭比时装舞台，女孩们可以参加真正的时装秀。芭比咖啡厅设在顶楼。

在零售空间里，设计师游刃于玩偶、女孩和女人之间，通过将现实与虚幻融合在一起，保持有趣和活泼的元素，增强了品牌的青春感，彰显了对无年龄界限的少女装扮的崇尚。在这个积极向上的芭比世界里，一切皆有可能。

2

1. A lively and flexible display area
2. The pink decorations are elegant and warm
3. The spiral staircase is flanked with Barbies
4. The fresh and lovely bathroom design
5. The hall is gorgeous
6. The pink escalator brings customers away from the busy daily life

1.活泼灵动的展示区
2.粉红色的装饰优美温馨
3.螺旋楼梯两侧被芭比娃娃包围
4.清新可爱的洗手间设计
5.大厅设计美轮美奂
6.粉红色的自动扶梯带顾客远离喧嚣

within the store is a three-storey spiral staircase enclosed by eight hundred Barbie dolls. The staircase and the dolls are the core of the store; everything literally revolves around Barbie.

The staircase links the three retail floors: the women's floor (women's fashion, couture, cosmetics and accessories), the doll floor (dolls, designer doll gallery, doll accessories, books), the Barbie Design Centre, where girls design their own Barbie on this floor, the girls floor (girls fashion, shoes and accessories). The Barbie Fashion Stage, where girls take part in a real runway show, is also on this floor. The Barbie Café is on the top floor.

Throughout the retail areas, the designers played with the scale differences between dolls, girls and women. They reinforced the feeling of youth and the possibilities of an unapologetically girlish outlook (regardless of age) by mixing reality and fantasy and keeping play and fun at the forefront – to create a space where optimism and possibility reign supreme as expressions of core Barbie attributes.

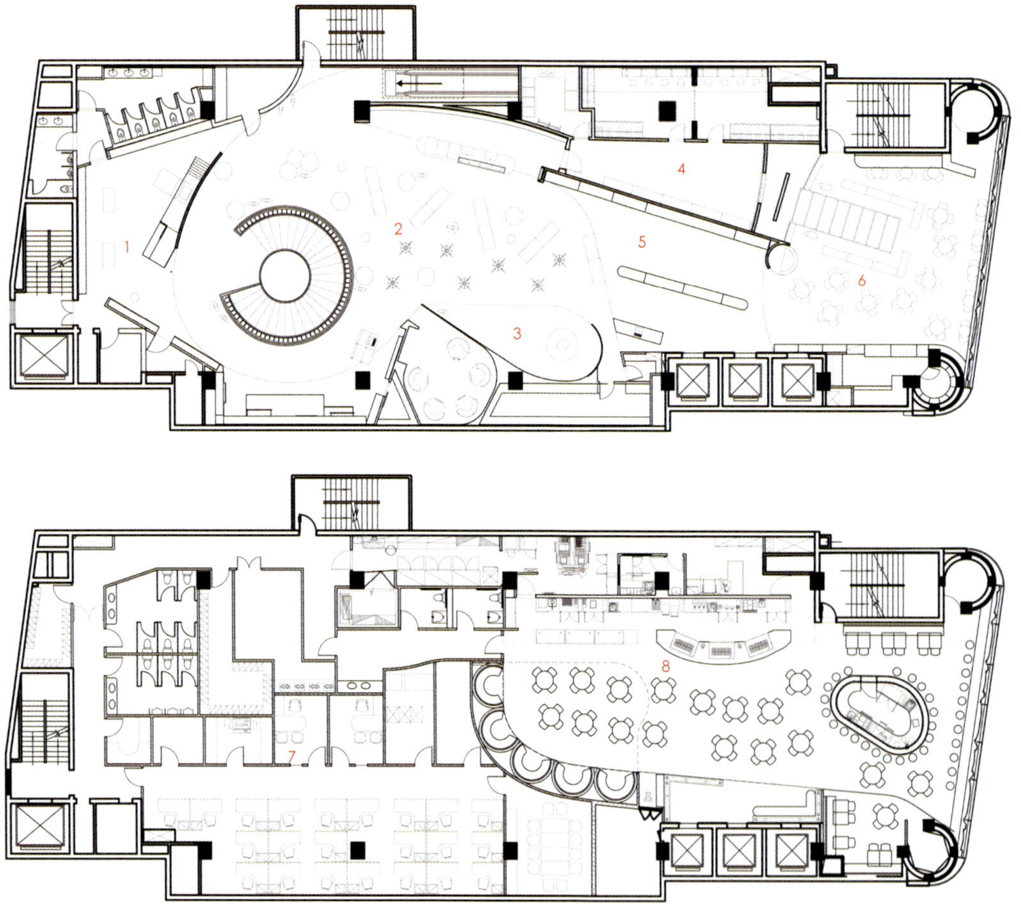

1. Home
2. Girl's fashion
3. Shoes
4. Fashion runway
5. Giftables
6. Café
7. Offices
8. Restaurant

1.前台
2.女孩时装
3.鞋
4.时装秀场
5.礼物
6.咖啡厅
7.办公室
8.餐厅

3

4

5

6

Fantasy World

幻想世界

Location:
Kuwait City, Kuwait

Designer:
JGA

Photographer:
Fantasy World

Completion date:
2010

项目地点：
科威特 科威特城

设计师：
JGA设计公司

摄影师：
幻想世界

完成时间：
2010

The store features fun, multi-coloured round ceiling elements that match the distinctive department colours, making it easier and more exciting for guests to navigate their way through the experience. The storefront offers a changing display of colour and patterns visibly different each time when the customer visits. Located at the storefront, the prominent positioning of Freddy, the brand's iconic toy soldier, serves as a welcoming gesture. Passersby are drawn to the store by its rich colour palette and internally illuminated ceiling discs. Bold graphic striping of the store's feature walls were accented by graphically rich floor materials; at the same time appearing simply delightful and high technology. The brilliant high gloss white fixturing contrasts the use of the colour-zoned departments, while positioning the toys and its packaging as the hero.

The use of an iconic identification system, reinforced by departmental colour blocking defines merchandising zones. Background paper and end-cap graphics create an intuitive, easy-to-follow directional system for either children or their time-pressed parents. The store's simple perimeter racetrack leads shoppers into a series of departmental valleys to offer the store intimacy. Uniquely designed impulse areas easily accessed by customers allow the store to feature products either by price point or end use. Merchandise displays with the price point, density and easy

店铺设计的特色在于有趣的多色圆形天花板吊顶，每个区域的颜色都不同，便于顾客寻找到自己所需的商品。店面的色彩不断变换，让顾客每次都有不同的体验。店门口的卡通玩具士兵弗雷德充当了门童的角色。丰富的色彩模式和发光顶棚吸引着行人的目光。店铺墙壁上的图形条纹在丰富的地板材料的映衬下格外显眼，既简单可爱又彰显了高科技特征。闪耀的高光白色展架与各个彩色区域形成了鲜明对比，让玩具和包装品成为了主角。

图标识别系统的应用，配合着各个区域的色块，划分了零售区域。背景画和图形标志共同形成了通俗易懂的导航系统，为孩子和时间紧迫的家长提供了方便。店内简单的环形走道引领顾客到达各个部门进行购物。特别设计的即兴购买区让顾客可以轻易获得店铺所展示的商品。根据价格、商品密度和易获得程度所进行的商品展示为顾客提供了更加高效的购物体验。每个商品部门都有大型屏幕，用来播放电影、卡通和其他广告讯息。

从店门口便可以看见幻想世界内部大胆而充满活力的色调和内置照明的圆形碟片，整个空间汇聚了不同色彩和造型，强化了“脱离现实世界”的主题。全息焦点元素保证了店铺内外色彩的一致性。

Amazing Minds
RADAR
RADAR SUPPORT
WITH STAINLESS STEEL CORBONIDE
PILOT PORTHOLE
REAR THRUST
Amazing Minds
SOLAR SYSTEM
EARTH MOON
QUARTZ MINING
GOLD MINING
VOLCANO MAKING KIT
BRACHIOSAURUS
STEGOSAURUS
TRICERATOPS
TYRANNOSAURUS REX
dig a dino Dinosaur
reflector telescope
télescope réflecteur

1. The colourful entrance
2. The toy area
3. Different lightings represent different zones
4. The change of colours provides customers with rich experience
5. The children bicycle area
6. Different areas gather different colours and forms
7. The lovely corner of the display area

1.缤纷多彩的入口
2.玩具展示区
3.不同的灯带代表着不同的区域
4.店内色彩的变化给顾客丰富的体验感受
5.儿童车展示区
6.各个空间汇聚了不同色彩和造型
7.展示区一角设计得活泼可爱

accessibility lends to an effective consumer-centric experience. Large monitors are highlighted in each department allowing for movies, cartoons, and other promotional messages to be broadcasted either on an overall or specific store basis.

The bold and vibrant colour palette, along with a brilliant array of internally illuminated fabric-stretched discs creates a playful constellation of colour, shape and elevation visible from the store's main entrance that reinforces the "out of this world" theme. The holographic perimeter focal element provides a perspective of continuity from the exterior colour morphing to the inside of the store.

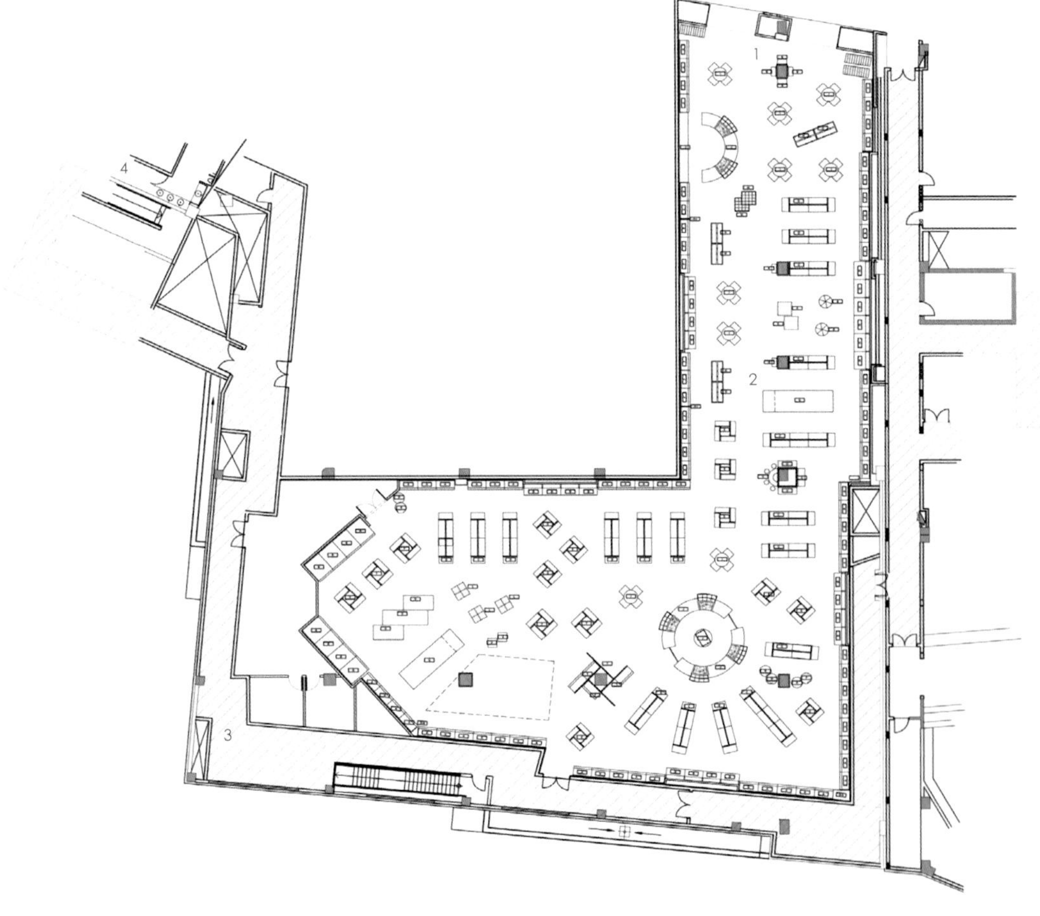

1. Entrance 1.入口
2. Display area 2.展示区
3. Mech. shaft 3.机械室
4. Washroom 4.洗手间

GIRLS
GIRLS
Early Years
Early Years
Early Years
Girls
Boys
KIDS
THE
BEST
KIDS
DESERVE
3

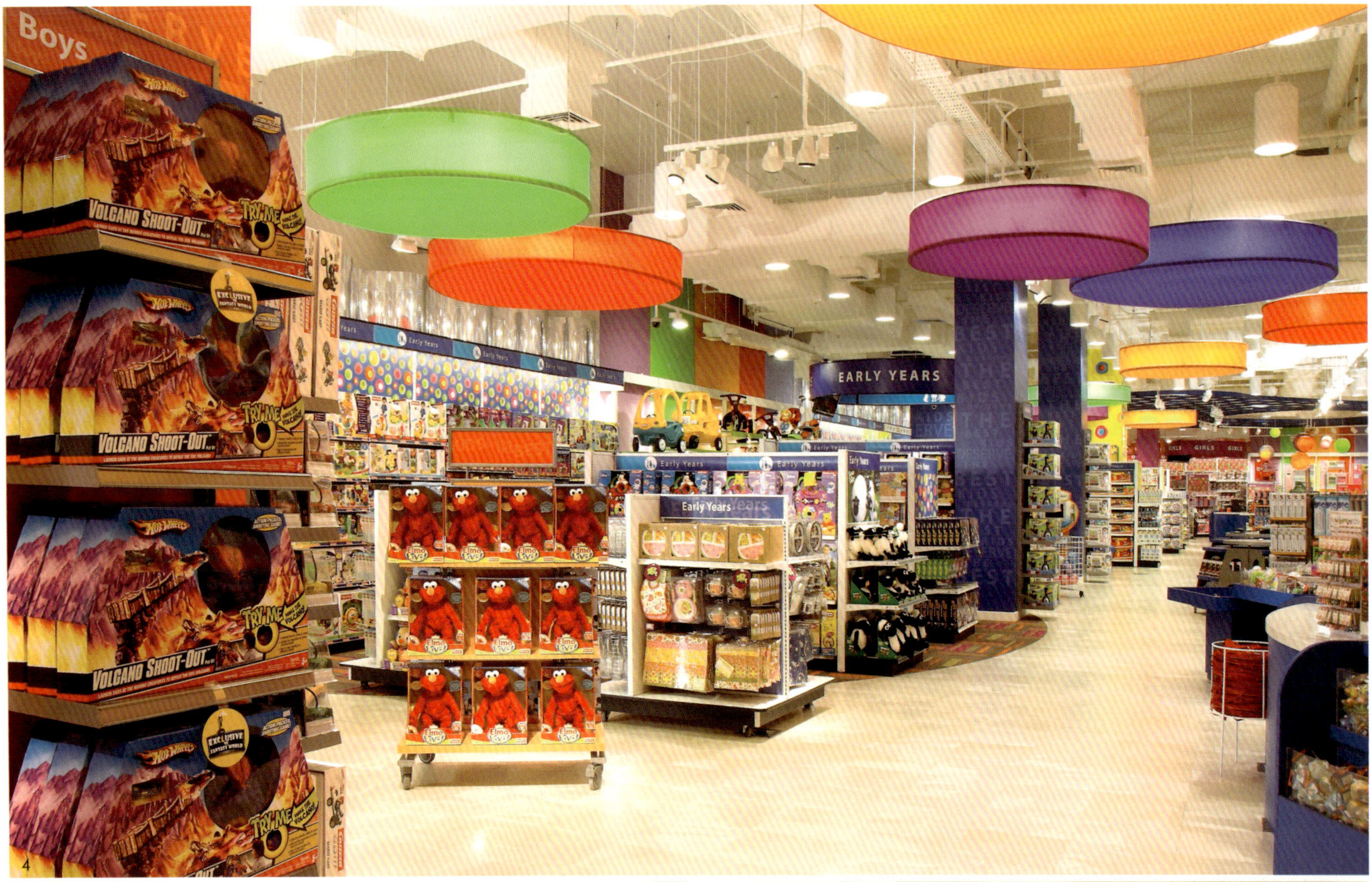

4

5

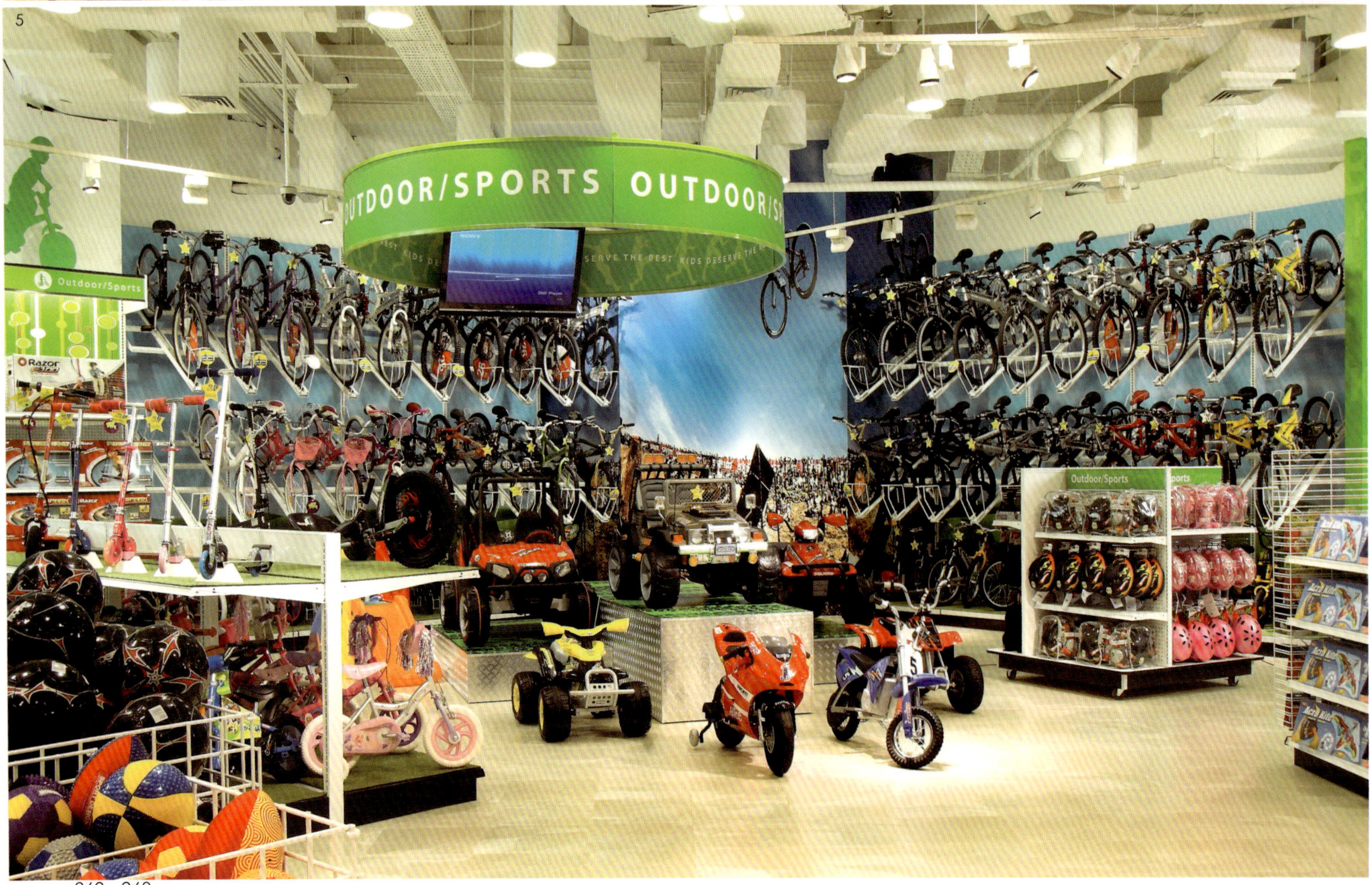

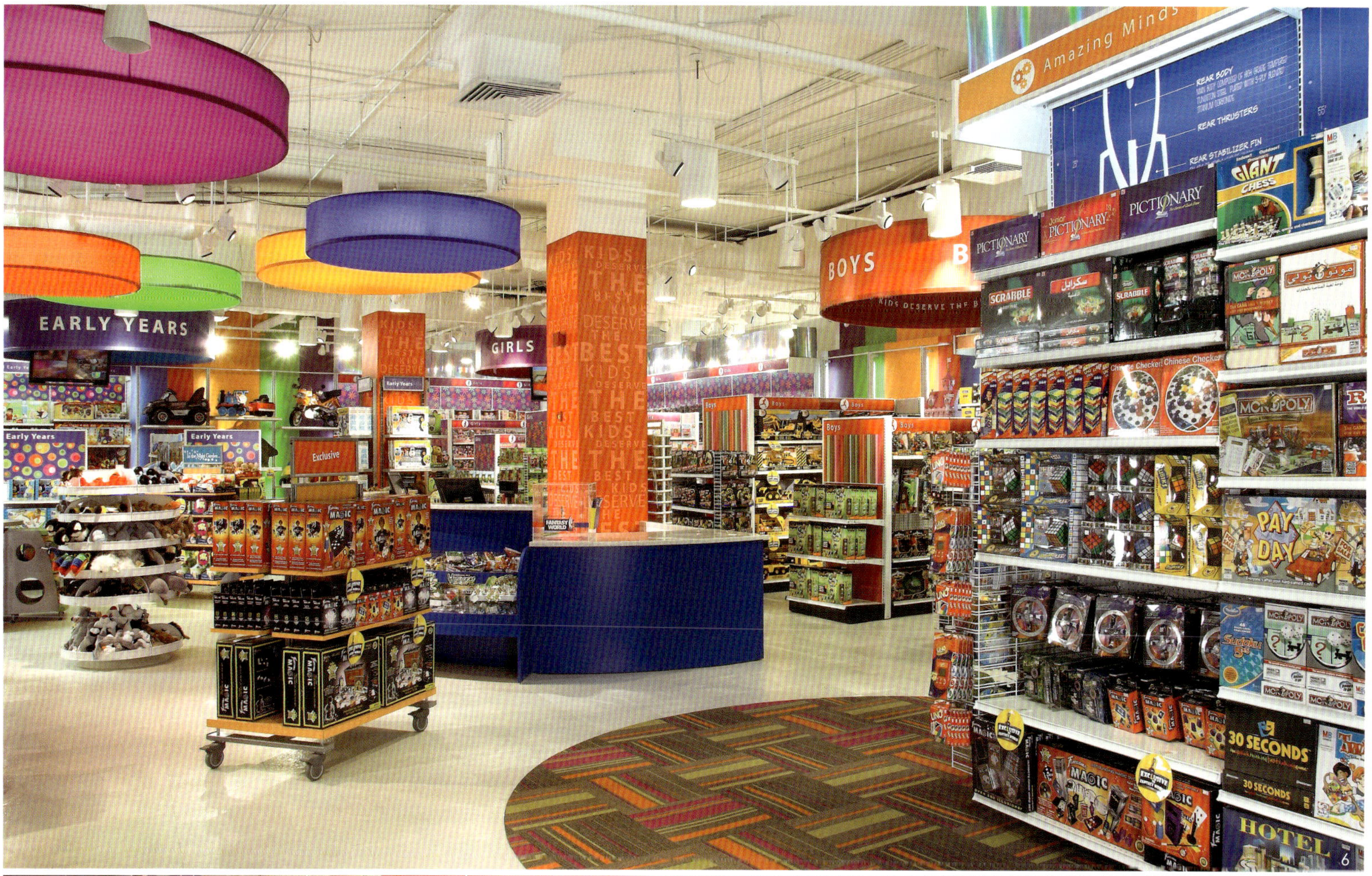
Amazing Minds
EARLY YEARS
GIRLS
BOYS
Exclusive
PICTIONARY
SCRABBLE
MONOPOLY
PAY DAY
30 SECONDS
HOTEL

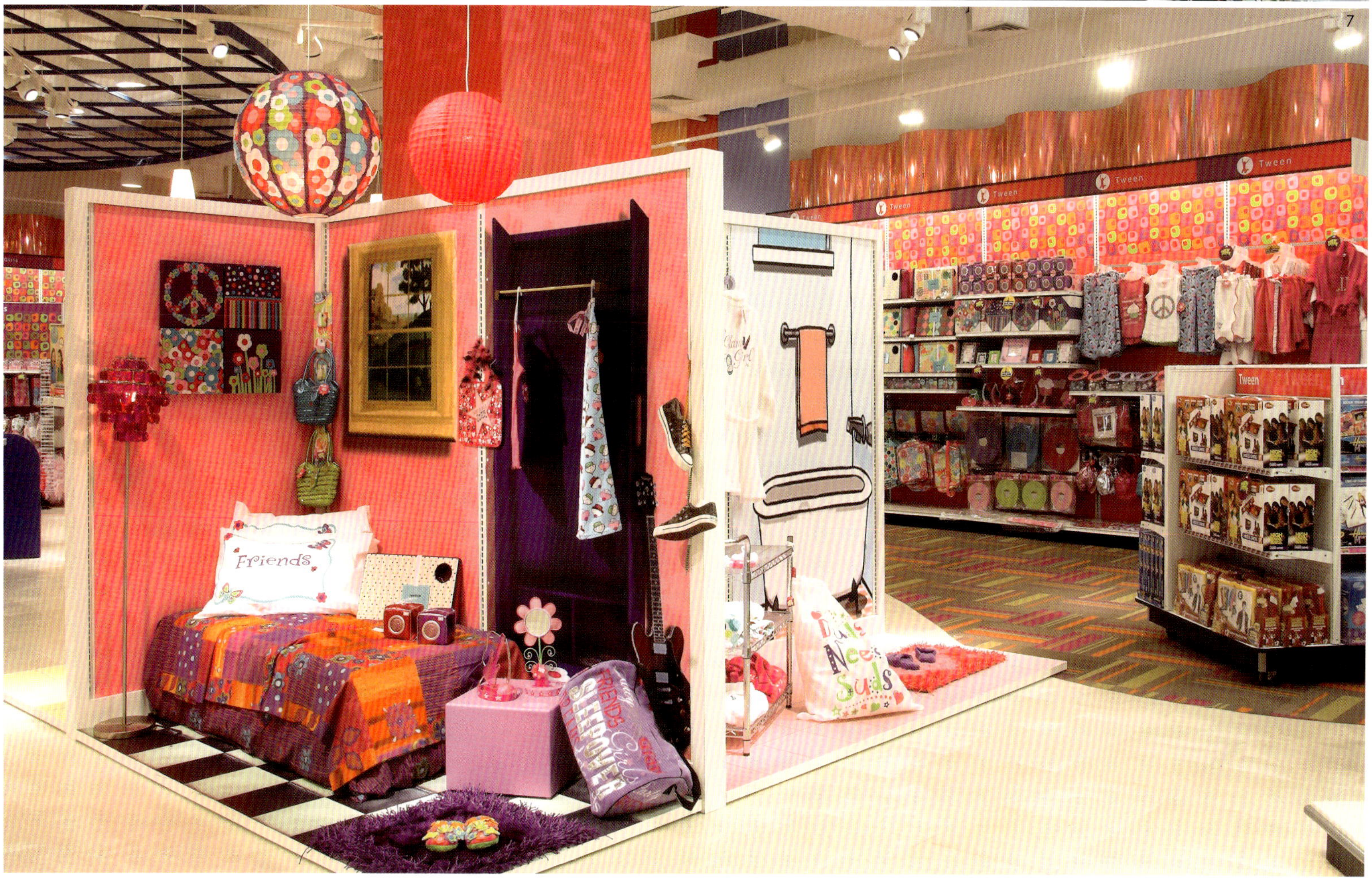
Friends
Tween

1

IMAGINARIUM in Barcelona

巴塞罗那IMAGINARIUM

Location:
Barcelona, Spain

Designer:
SAKO Architects

Photographer:
Marta Pons Miralbell

Completion date:
2007

项目地点：
西班牙 巴塞罗那

设计师：
迫庆建筑事务所

摄影师：
玛尔塔·彭斯·米拉贝尔

完成时间：
2007

"IMAGINARIUM in Barcelona" is a flagship shop of a well-known Spanish children's brand called "IMAGINARIUM", and located across the Passeig de Gracia from "Casa Mila". The shop occupies three floors (B1F-2F) of a narrow building 6 metres wide by 28 metres long.

Since the shop deals in 23 different series of commodity, the designers attempted to create a space that has not only many independent rooms match to each series, but also continuity as a whole. The designers were inspired by a space characteristic of "Cave", which is that there is no viewpoint people can perceive whole space but instead different scenes are explored continuously as people go deeply into the space.

Many of their goods and packages are in primary colour, therefore the designers filled whole space with rainbow colour to set off the commodities and create a cheerful atmosphere. Colour on the wall gradually changes in vertical direction. Each of six layers of display shelves draw their own smooth curved line throughout the shop, which is like undulations of natural land drawn by contour lines. In some places, display shelf is flying out just like "tongue" and becomes special shelf for promotional commodity, area sign, table, chair, counter, book shelf and partition and so on. Those "tongue" create distinctive areas in the shop then generate variety of activities around

巴塞罗那IMAGINARIUM是儿童品牌IMAGINARIUM的旗舰店，位于Passeig de Gracia大街Casa Mila对面。是一个6米宽、28米进深的细长空间，共有三层（B1F、1F、2F）。

该店铺经营的商品多达23种，因此在设计上既要考虑到多样空间的需求来对应不同商品，也必须注意空间的整体性。洞窟是正好符合要求的元素，一眼望去并不能将全部景象尽收眼底，而是随着空间的深入，整体景象慢慢连续的展现在眼前。

商品系列多为原色。因此设计师在空间中使用彩虹来突出商品，并营造热烈的气氛。彩虹的颜色也在垂直方向随高度增加而逐渐发生变化。6层的陈列柜，每层都强调了圆滑的曲线。好像"等高线"一样再现了自然的地形。一些像"舌头"一样突出的陈列柜可以活用于促销产品的展台、区域标志、桌子、椅子、柜台、书架或者隔板等各种功能性需求。由于展架每层形状不同，因此各种商品被分散的分布在三维空间中。由上而下俯视时，可以看到各种颜色的展架好像"等色线"，勾画出了商品的分布，也同时强调了"空间的应力"。

从入口开始，店铺充分为成人及儿童配置了各种大小配套的设施，以此来强调IMAGINARIUM品牌同等对待成人及儿童的理念。儿童在洞窟中玩耍远远超越了购物的概念，而是引导并刺激了孩子的探险欲。

WALKIDS

1. The entrance with rainbow forms
2. The stack display shelf
3. The relaxing area
4. The curves of showcases look like natural topography
5. The products are arranged in this 3D space
6. The background wall with rainbow colours on it

1.彩虹造型的入口
2.层叠的商品展示架
3.店内休息区
4.陈列柜圆滑的曲线仿佛自然的地形
5.商品均匀的分布在三维空间里
6.彩虹色的背景墙

them. Since each layer has different shape, activities are distributed irregularly in three-dimensional way. So once you over look from right above with all layers translucently overlaid, colour lines show the distribution of activities, or "space stress".

As represented by a pair of doors at the entrance, goods for parents are displayed right next to those for children, which embody IMAGINARIUM's philosophy of treating children as adults. For children the experience of exploring in "Rainbow Cave" will exceed that of usual shopping and stimulates their native curiosity.

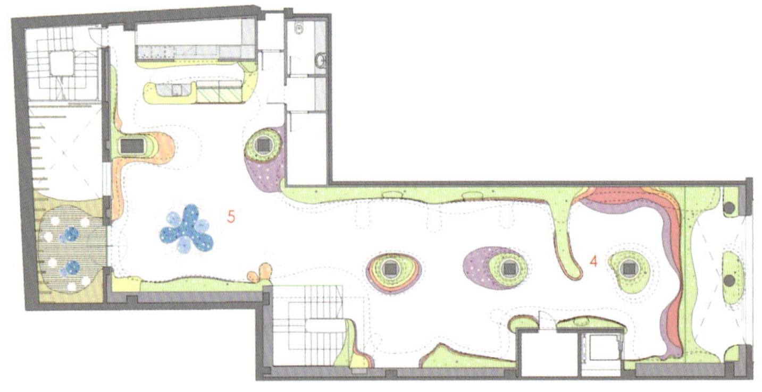

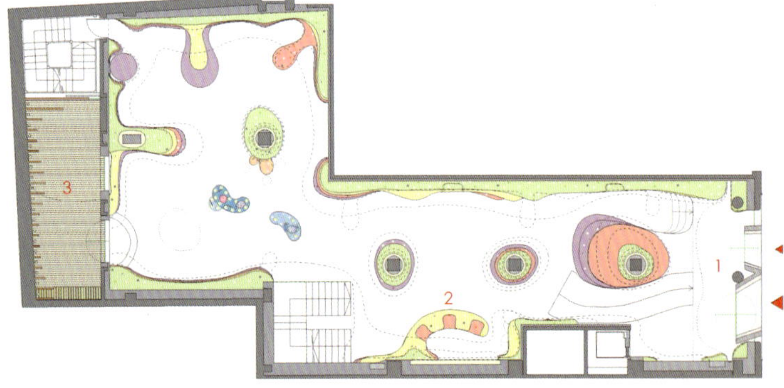

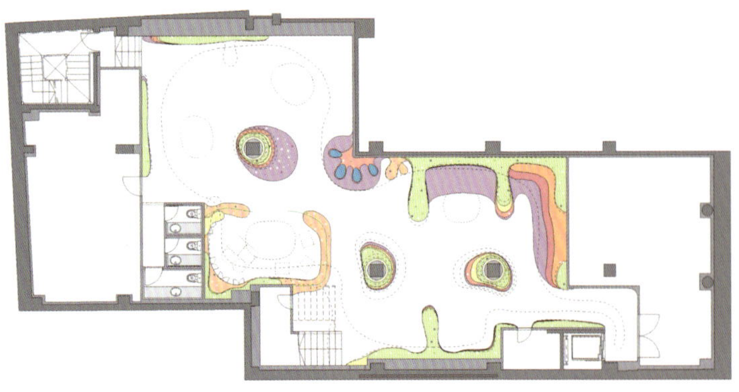

1. Entrance 1.入口
2. Cashier 2.收银台
3. Terrace 3.平台
4. Bookshop 4.书店
5. Café 5.咖啡厅

4

5

6

Index
索引

图书在版编目（CIP）数据

体验空间 / 于萍编 ；常文心译 . -- 沈阳：辽宁科学技术出版社，2011.7
ISBN 978-7-5381-6940-9

Ⅰ. ①体… Ⅱ. ①于… ②常… Ⅲ. ①商场－建筑设计－作品集－中国－现代－汉语、英语 Ⅳ. ①TU247.2

中国版本图书馆CIP数据核字（2011）第076294号

出版发行：辽宁科学技术出版社
（地址：沈阳市和平区十一纬路29号 邮编：110003）
印 刷 者：利丰雅高印刷（深圳）有限公司
经 销 者：各地新华书店
幅面尺寸：230mm × 290mm
印 张：34
插 页：4
字 数：50千字
印 数：1~2000
出版时间：2011年 7 月第 1 版
印刷时间：2011年 7 月第 1 次印刷
责任编辑：陈慈良
封面设计：迟 海
版式设计：迟 海
责任校对：周 文

书 号：ISBN 978-7-5381-6940-9
定 价：268.00元

联系电话：024-23284360
邮购热线：024-23284502
E-mail: lnkjc@126.com
http://www.lnkj.com.cn
本书网址：www.lnkj.cn/uri.sh/6940